TRUE GENIUS

TRUE
GENIUS

The Life and Work of
RICHARD GARWIN
The most influential scientist
you've never heard of

JOEL N. SHURKIN

Prometheus Books

59 John Glenn Drive
Amherst, New York 14228

Published 2017 by Prometheus Books

Interviews of Richard Garwin by Finn Aaserud courtesy of Niels Bohr Library & Archives, American Institute of Physics, College Park, MD.

Cover image © Abigail Dunlap
Cover design by Jeff Schaller
Cover design © Prometheus Books

Inquiries should be addressed to
Prometheus Books
59 John Glenn Drive • Amherst, New York 14228
VOICE: 716–691–0133 • FAX: 716–691–0137
WWW.PROMETHEUSBOOKS.COM

21 20 19 18 17   5 4 3 2 1

Library of Congress Cataloging-in-Publication Data

Names: Shurkin, Joel N., 1938- author.
Title: True genius : the life and work of Richard Garwin, the most influential scientist
    you've never heard of / by Joel N. Shurkin.
Description: Amherst, New York : Prometheus Books, [2017] | Includes bibliographical
    references and index.
Identifiers: LCCN 2016032730 (print) | LCCN 2016035282 (ebook) |
    ISBN 9781633882232 (hardcover) | ISBN 1633882233 (hardcover) |
    ISBN 9781633882249 (ebook)
Subjects: LCSH: Garwin, Richard L. | Physicists—United States—Biography.
Classification: LCC QC16.G38 S58 2017 (print) | LCC QC16.G38 (ebook) |
    DDC 530.092 [B] —dc23
LC record available at https://lccn.loc.gov/2016032730

Printed in the United States of America

For my grandsons Rafael, Noam, Saadia, and Aaron Shurkin,
and my daughter Hannah.

CONTENTS

ACKNOWLEDGMENTS

First and foremost, I must thank the Alfred P. Sloan Foundation whose grant made this all possible.

And, of course, Dick Garwin himself, whose cooperation was sometimes almost too much. Many mornings we had what I began to call "Garwin document dumps." I would open my computer and Dick had sent more documents. The database on my Scrivener file was enormous, containing more information than I could ever possibly use. He did a great deal of the research for me. Now eighty-eight years old at this writing, he has an astonishing memory. He can recall names, dates, titles, etc. Only he could keep track of the acronyms for all of the government agencies, most of which he worked with. He was almost never wrong. He was a biographer's dream.

The book was the idea of Tony Fainberg of the Institute for Defense Analysis, who also raised the money to help me get started, who read some of the manuscript, and who is a true *mensch*.

This is my third biography, and each had different problems. My first, about the psychologist Lewis Terman, relied almost entirely on written material. Terman had died years before, and there was no one still alive who interacted with him. The second was on William Shockley, the physicist who had died only ten years earlier, and there were many people I could talk to. They were all very willing, mostly to tell me how much they hated him.

Dick Garwin is the last of what may be seen as the golden age of physics in the middle of the twentieth century. He is one of the very few people alive who knew and worked with Enrico Fermi (who called him the best graduate student he ever had) and dozens of Nobel laureates, legends in their field. Except for Sid Drell, now in his nineties, he is the only one left, so Garwin's dumps were particularly valued on this end.

Also I had writer Daniel Ford's work. Ford had begun a biography of Garwin in the 1990s and then had to abandon the project. Meanwhile, he had done a score of interviews with people who had worked with Garwin, many of whom have since passed on, and I would never have had access to their experience otherwise. He gave me permission to use those interviews, and I have been careful to give him credit, either in the text or the references.

A special bow to Grace McClintock, my assistant in the last days. Grace is a neighbor, my daughter's best friend, and sometimes acts as my adjunct daughter. She conveniently graduated from Johns Hopkins with a degree in physics just in time to help me out. She was my fact checker and was in charge of the references and bibliographic material, and provided insight and grace to my office.

Full disclosure: the manuscript was seen by Garwin with the request that he check for accuracy. I am especially grateful in places where I wasn't sure I understood what was happening. See the chapter on muons. He exercised no editorial control nor did he ask for any. The book is entirely mine, and if there are errors, I'm responsible. That also is true of the opinions expressed unless they are attributed.

The book was written using Scrivener software, and in the end we invoked Mellel and Google Docs for collaboration and formatting.

Thanks to my cousins Bunny and Paul Lichtenstein for room and board in Santa Fe. It was Paul who told me about 109 East Palace Street, the secret door to Los Alamos, and Jennel Conant's wonderful book about it. Also Ann Finkbeiner's fine book *The JASONS* was most useful.

Several people read portions to make sure I made some sense, including Tony Fainberg, my son Michael, who is a political scientist and historian at the RAND Corp., and Alan Cohen, gentleman, lawyer, and historian. I asked Alan to read the chapter on Vietnam because of his interest in history, not knowing he was a Vietnam vet.

And, of course, my wife, Carol Howard, who despite her illness was patient and kind when I was distracted by the writing.

INTRODUCTION

American physicists have a "Garwin joke" they tell about their esteemed colleague, Richard Garwin.

Somehow, Garwin and two other men are arrested during the French Revolution and sentenced to the guillotine.

The first prisoner is escorted up to the deadly machine and then tied to the execution platform, his head in the wooden restraint. The blade goes up. The executioner releases the rope, and the blade stops an inch above the prisoner's neck. Since the law forbids someone from going through the execution process twice, the executioner must let him free.

The second man is brought to the guillotine, and the blade goes up, and slides down. And it too stops an inch above the prisoner's neck. He is also set free.

Finally, Garwin is brought up. He is tied to the platform. Just before the executioner releases the blade, Garwin looks up and says, "I think I know what your problem is."

Credit for whom did what in the development of the hydrogen bomb is not easily established. Whoever did it changed the world. Since then we have lived on the precipice. Much of the work still is classified, although scientists in many countries have figured most of the secrets out by now, either by themselves or through collaboration, or espionage. The United States, Britain, France, China, and Russia have thousands of nuclear or thermonuclear weapons; Israel probably has several dozen. India and Pakistan have fission bombs and may be working on thermonuclear weapons. Almost all of the US nuclear arsenal consists of variations on hydrogen bombs in all shapes and sizes, ranging from small bombs that could wipe out a company of attacking soldiers to weapons of genocide. The US even explored nuclear artillery but stopped in 1991. At one time thermonuclear hand grenades and mortars were suggested, although finding anyone foolish enough to lob one might have been a problem. The bigger devices have no military use—they exist to terrify would-be adversaries into not attacking.

Additionally, the personalities involved were very human. Edward Teller spent most of his life known as the "Father of the Hydrogen Bomb." Depending on who is telling the story—and what the person thought of Teller—Teller either came up with the idea that made the bomb feasible or he did not. In the latter case, then Stanislaw Ulam, whom he disliked, did. In many circles, the bomb is known as the Teller-Ulam device, or the Teller-Ulam Super, with the names always in alphabetical order to avoid controversy. Others made contributions, including Enrico Fermi and Hans Bethe, but Teller insisted he, not Ulam, did the key work. Others said Teller couldn't have done it without Ulam. Since much of what they did is still classified, it may not be straightened out for years—if ever. Somehow the two of them worked it out, coming through with the concept that erased doubts that such a bomb could be built. It is likely they could not have done it without each other.

But in fact, Garwin, a then twenty-three year-old postgraduate student, took their concept and came up with the design that led to the bomb, although he never got public credit. His role was completely unknown outside of a small group in Los Alamos.

This is the story of how that happened and how that event changed the world and the life of one of America's most brilliant and political of scientists. Some of what you are about to read has never been published before. The story has everything to do with both why we are still alive as a species and the declining role science plays in government. A prodigious polymath, Garwin has affected almost everyone's life with his more benign inventions, from air traffic control systems to the first laser printer, and he was a pioneer in the development of touch screens. But of the bomb, he once said, "If I had a magic wand, I would make it go away."

He has dedicated his life to producing that wand.

"In Japan they declare certain people national monuments, and Dick is a national monument," said a colleague.

CHAPTER ONE
THE TINKERERS

In the summer of 1950, Richard Garwin, a young PhD in physics, and his wife, Lois, went to Los Alamos, New Mexico, with their new son, Jeffrey. They were immediately awed.

They were taken by the scent of the place, the mixture of dust and pine, and by the Jemez Mountains and the Sangre de Cristo range looming twenty miles away. On days when Los Alamos was under a dark cloud, the mountains could be brightly sunlit. The physicist in Garwin noted that when "the cloud covered most of the path over the valley that would otherwise have dimmed the contrast of the view by contributing a background of scattered light."[1] They had driven to Santa Fe and then across the sand-colored landscape, up a perilous mesa road to the place where scientists had gathered to invent the bomb that destroyed two cities in Japan and helped end the greatest war in history.

Returning to Los Alamos for the summer of 1951, Garwin had just spent the winter in Korea and Japan at the behest of the US Air Force because the air force had set up the Tactical Air Command and wanted to know what kinds of technologies and laboratories would be most useful in the war then embroiling the Korean peninsula. Garwin had been a graduate student of the great Enrico Fermi at the University of Chicago and was already becoming well known by the defense establishment, a relationship—sometimes smooth, sometimes corrugated—that would last the rest of his long life. He found the work at Los Alamos interesting. Just as important, room and board at Los Alamos was free, and he did not have a full-time job. Instructors at the University of Chicago were paid $4,700 for nine months' work, "which was not really enough to live on," he said.[2]

The first thing he did was to make use of his new security clearance (granted because of his work at the Argonne National Laboratory) so he could read all the secret papers in the library on nuclear weapons.[3]

One day he was chatting with Edward Teller, whom he knew from Chicago. Garwin, like everyone else in physics, knew that Teller was fascinated almost to the point of obsession with the possibility of a hydrogen bomb, a weapon vastly more destructive than the fission bomb developed at Los Alamos. Everyone also knew that Teller did not actually know how to build one.

Garwin asked Teller what was new and how he could be of assistance.

He could do a great deal, it turned out. . . .

Richard Garwin, appropriately enough, came from a family of tinkerers. Immigrant tinkerers.

The original family name was Gawronski, and his paternal grandfather came from Riga—or so Garwin thinks. His paternal grandfather immigrated to Chicago and opened a shoe store. Garwin's father, Rubie (who later changed his name to Robert), was born there in 1898, the third of four brothers. When Robert was seven years old, his father was murdered by his partner, leaving Robert's mother alone with four boys. She moved to Cleveland because the city's large and thriving Jewish community had an orphanage, the Jewish Orphan Asylum, where she put two of the boys, including Robert, because she could not support four children alone. The orphanage had originally been built for the children of Jewish Civil War veterans. It was not an entirely pleasant place. The two brothers spent all their school years there.[4] In 1920, the entire family changed their names to Garwin.

In 1921, Robert got an engineering degree from the Case School of Applied Science (which later merged to become Case Western Reserve). None of the other Garwin boys went to college. Robert served in the US Army during the First World War but never left Cleveland. He never became an engineer either, Garwin said, probably because of anti-Semitism in the profession. Instead, Robert became a science teacher at East Technical High School, specializing in teaching electricity by day and becoming a movie projectionist at night.[5]

Garwin's mother, Leona Schwartz, was a "greenhorn," arriving in 1912 from Hungary where she was born in 1900. Leona was the second of twelve children—nine of whom survived to adulthood. She went to work in a department store instead of high school, and married Robert Garwin in 1923. She worked as a legal secretary.[6]

Richard Garwin was born April 19, 1928.

He was a happy, healthy child, and his aunt Margie, puzzled by his mental development, gave him arithmetic tests ("three times six, plus seven, times two, minus fifteen, divide by . . ."), which he handled with aplomb.

"And I was so amazed at his vocabulary and what he knew—at one. Actually, after a while, when I had my children, I realized how advanced he had been at one year old. At that time we didn't know," his aunt remembered. "He had an amazing vocabulary."[7]

His father, Robert, spoke to him more as if Garwin were an intelligent adult than a child, explaining things mechanically.

"I was also curious about the natural and engineering world, and proved to be good at tools and books," Garwin said. "I was notorious for taking things apart and—mostly—putting them back together again."[8] He helped his father build Mahjong sets, a tile game that for decades was a fad with Jewish women despite its Chinese origin. The sets had pegs that held colored washers, which substituted for money, and fake ivory tiles that clicked when they were played.

When Margie's family got a television set, Garwin and his brother, Edward, took it apart to see what was inside and then put it back together again. For his twelfth birthday, he asked for a calculus textbook, Margie said.[9]

In 1940, Robert realized their house was "underwater," meaning they owed more than the house was worth. So, like many others, they walked away from their home and somehow got a mortgage for a new, two-family house in the largely Jewish suburb of University Heights. Robert told the builder to double the size of a two-car garage even though they only had one car. He built a workshop in the space the car didn't need.

Projectionists were highly skilled, unionized workers. But when sound movies threatened the industry many, including Robert, had to learn the new technology, which included adjusting and maintaining sound-on-film speakers and amplifiers. The local school board had a rule that its teachers could not hold two jobs—it was the Great Depression, and the feeling was that as many people as possible should have jobs—so he quit teaching and became a full-time projectionist and then opened a motion picture equipment and sound repair business in a twenty-foot extension to the Washington Boulevard garage he built when they bought the house. His brother Joe joined the business, Gartec Theater Equipment, and eventually it grew big enough for them to hire employees. Robert gave classes to other projectionists in the city for many years, and ran the business until his death in 1979.

The family, cousins, aunts, uncles, came and went. The Cleveland house was crowded. Garwin's aunt Irene lived with her husband and daughters on the first floor until they were divorced; Garwin, his parents, and Garwin's brother Edward lived on the second, and his father had a workshop to himself in the attic. He and his father were an engineering team, throwing together projects.

Garwin was not the only member of the family to benefit from Robert's intellect and talent. Edward also went on to become a distinguished physicist, with a doctoral degree from the University of Chicago, studying under Valentine Telegdi, who would become one of Garwin's friends and competitors. Edward spent most of his career at the Stanford Linear Accelerator in Palo Alto, California, helping to build the two-mile-long electron accelerator and the Stanford Positron Electron Accelerating Ring. Edward died in 2008.

Remembering his father, Garwin said, "He was a very capable person, and I liked it because he did interesting things." To call someone "capable" is the highest compliment one scientist has for another. "He was very good with his hands and thinking things through, and he wanted to teach me everything he knew. I did not resist."[10]

Both families shared the basement, with its coal pile, the furnace, and washtubs. Garwin and his father built a darkroom in the basement, next to the coalbin.

"Of course, by the time I was twelve years old, I was very much interested in science already. He [Robert] had a lot of books around the house, engineering manuals and so on, which I read at an early age. I was a good student. My handwriting was terrible, but otherwise I was a good student. I learned to type at the age of seven or eight, because otherwise the teachers wouldn't accept my work."[11]

"I would read the *Mechanical Engineer's Handbook* or the *Electrical Engineer's Handbook*," he wrote later, "and I had read much of the encyclopedia in grade school, but without mastering it all. From 1940 to 1947, I helped my father in his sound equipment business, with tasks ranging from splicing film and cleaning equipment to building amplifiers, and the like."[12] He analyzed the cathode "follower" vacuum tube circuit, an amplifier used in sound projection, to study how it and the other tubes used in amplification worked.

"But throughout my youth I was shy and not good at sports," he wrote once.[13] He was the kid who was chosen last for a baseball team. He did like to swim.

"My father had a .22-caliber single-shot rifle, and a .22-caliber automatic pistol (both for target practice—not for hunting), with which I became a good shot."[14]

He also played with chemistry sets. "I am ashamed to say that I risked my own life and that of my family by producing some explosives, knowing full well that Sir Humphrey Davy had lost several fingers to one of them."[15] (The chemical pioneer Humphrey Davy damaged his right eye in an experiment with nitrogen trichloride. It was Pierre Louis Dulong who lost his fingers—and an eye.)

Garwin also acquired a knack for glassblowing—strictly utilitarian. He lined a wooden bench, which they used for glassblowing, with marble plates to keep it from catching fire. They bought professional glassblowing equipment—ribbon burners, controllable torches—and hooked up an old vacuum cleaner to pump air and natural gas into what became the furnace. He wasn't making art; he was making chemistry equipment, tubes and condensers.

He also signed up for the Westinghouse Talent Search, a nationwide contest that rewarded young people for designing science experiments. Garwin's involved measuring the voltage required to decompose water by passing a current through it under pressure. He set up a heavy glass shield for protection, which came in handy when the experiment exploded with a deafening bang. It didn't matter; the teacher forgot to turn in the report on time.

Garwin was assigned to what was called a "major work program" at school, something like a gifted student's program in modern jargon. Because he skipped two grades, he graduated at the age of sixteen from Cleveland Heights High School.

He had decided on physics as a career. He applied to the University of Chicago and Allegheny College, but the Case School of Applied Science offered him a half scholarship, and family finances being what they were, Case won. He lived at home, taking the bus to school. He went to classes and immediately returned home, taking no part in the extracurricular activities or social life the school offered. He did not have time nor interest, and, with the war looming, everyone was in a rush. He finished in three years, again acceleration as the result of wartime needs. He got perfect grades in every course but one, and that one exception involved a registration mix-up.

Garwin had been a college sophomore on August 5, 1945, when the government announced that an atomic bomb had been dropped on Hiroshima. "I thought it was very interesting and a big achievement that one could obtain energy from the nucleus," he told interviewer W. Patrick McCray. "Of course anybody who had been following it would've known that, but I hadn't. I guess in 1945 I was already in college, but people didn't talk about nuclear energy. People who knew anything about it realized it was classified and most of those people weren't at the universities anymore anyhow. They'd gone into war work of one kind or another."[16]

"There was no television at the time—newsreels and photographs published in the newspapers," he wrote, "but the world was agog with this. . . . By then I did know something about atomic physics and a little bit about nuclear physics, so I could understand to some extent what had happened. Of course, I moved quickly to obtain a copy of the Smyth Report, published August 1945 by the War Department, which revealed what details could be told of the atomic bomb program (Manhattan Project) in the United States."[17]

He did have one distraction: one of his brother's friends, Howard Levy, spent considerable time at the Garwin home. Occasionally, Levy's sister Lois would call to see whether Howard was there and to ask someone in the Garwin home—often Garwin—to send him home. Their mother and Garwin's would later become friends, and Garwin knew Lois from junior high school.

Her father, Harry, was born in London, her mother, Bernice, in Providence, Rhode Island. Lois initially grew up in the small town of Willoughby, Ohio, a Cleveland suburb. Her father was a haberdasher until the Great Depression when he opened a grocery store in Cleveland. Lois worked there half of every day, going to school the other half, and also did the cooking for herself and her brother. She found some escape, mostly at the Cleveland Playhouse program. Paul Newman was a classmate.

"Well, he pursued me," she told Dan Ford in an interview in California years later about Richard's courtship. "I wouldn't say he was flirtatious, no. He was very serious. Very serious, in fact. I could tell that he would think of things to talk about ahead of time. But he did pursue me." He was aware she was a hall monitor, which meant he knew where to find her at school.[18]

Lois had two years of college at Flora Stone Mather, a women's division of Western Reserve University, and she and Garwin married on April 20, 1947. He was nineteen. They are still married sixty-nine years later.

He began a lifetime of inventions. One was an intense television projector, making use of a more powerful electron beam than in the standard television tube. He didn't complete it. He also tried to grow diamonds at relatively low temperatures, but that didn't work.

Garwin graduated from Case in 1947. He was given a Standard Oil of New Jersey scholarship to do graduate work at Chicago. A professor at Case told Garwin and his father one day at lunch that he would be out of his mind to turn down Chicago, which had the best physics department in the country, with Enrico Fermi and Edward Teller on faculty.[19]

"The University of Chicago was my choice because of the presence of a stellar faculty in physics and allied departments, including notably Enrico Fermi," Garwin recalled.[20]

"He was very reluctant to go to Chicago and leave me behind," Lois told Dan Ford. "Sounds—it's so juvenile. When we started having children, we used to discuss, what are we going to tell our children to keep them from marrying young? Not that we hadn't had a success, but we realized how lucky and how unusual it was to marry at such an early age and make a success of it. Well, times had changed so drastically by the time our children were twenty or whatever, none of them was interested in getting married. I mean, it was the farthest thing from their minds. So you see, the things you worry about are the things that never happen."[21]

Lois was in college in Cleveland until the Garwins moved to Chicago where she began work at Blue Cross of Chicago. In the summer of 1948, back in Cleveland, she went to work first at Blue Cross, and then Ohio Bell.[22]

They could support themselves with Lois's job and Garwin's scholarship, but the problem was finding a place to live. With the war over, millions of GIs took advantage of the GI Bill of Rights to go to college. Simultaneously, millions of people from the South, many of them African Americans, looking to escape poverty, were migrating north, and Chicago was a prime destination. The Garwins could only find temporary shelter and moved thirteen times in the first twelve months to eleven different places. Often that meant housesitting while residents were away. Sometimes they had to spend nights in neighborhood hotels through the year. Garwin described his graduate work as both a joy and a strain with the constant worry about where they were going to live.

Their first child, Jeffrey, was born on November 18, 1949. They eventually took over a one-room studio apartment with a fold-out bed on South Shore Drive and Seventy-Seventh Street, and eventually moved closer to campus on East Fiftieth Street, an apartment created by fellow physics graduate student Harold Agnew out of a wraparound porch. "This was heaven for us, and we were able to buy furniture and to have a place for Jeffrey's crib when he came home from the hospital."[23]

Garwin had another stressor. The routine called for a student to take two tests before he or she began working on a dissertation. After two years of classroom study, the student would take the Qualifying Exam the following year, and if the student passed, it would mean a master's degree and entry into the PhD program. After a year or two, the student would take the Basic Exam. If he or she passed that, research could begin that would lead to the doctorate. Garwin was too impatient to go through that—he was fairly confident he could pass either—and bargained with the physics department that he would jump to the Basic Exam, and if he did well, he would skip the Qualifying. If he failed, he would have to take the Qualifying before he could try the Basic again.

"I got the highest grade in the class. Some people actually failed and had to take over, including some people who had Nobel prizes who I will not mention. . . . It was a substantial gamble for us," he said later.[24]

No worry: he entered the university in September 1947, passed the Basic in the following April, and began working on his thesis. Still restless, he felt he needed to find a faculty member to act as a mentor and to turn him loose in the lab. He wanted to be close to the research and "perhaps make some contributions of my own."[25]

"I was getting itchy because I had nothing to do with my hands," he said.[26]

He decided to start at the top, and that was Enrico Fermi.

CHAPTER TWO

FERMI

In what was a golden age of physics, Enrico Fermi was simply known as the "Pope." In physics, he was infallible. Someone described him as a "darling of nature." He never read whole scientific articles; it was faster and more efficient to look at the abstract and figure the rest out himself.[1]

"He was an extremely nice person," Garwin said. "He was small. If you saw him walking down the street, you wouldn't know he wasn't the corner grocer. He had large, round brown eyes that were very distinctive. Fermi was proud of his vision. He said he could see better than other people, which he did, both literally and figuratively. He was very competitive athletically. . . . He was indomitable."[2]

He had a high-pitched, slightly nasal voice that sounded as if he was speaking slower than he actually was. He was, Garwin remembered, an excellent lecturer, clearly delineating what he was talking about on a huge blackboard. He was easy to follow in class, but when Garwin got back home and looked at his notes, he had trouble reproducing what Fermi had said. Fermi always took mental shortcuts to get to his conclusions.[3]

Fermi had been the discoverer of the weak force, one of the four forces of nature (gravity, the strong force that holds atoms together, and electromagnetism are the others). Adopting the theories of the physicist Wolfgang Pauli, Fermi, working in Rome with neutrons from 1934 to 1938, proved you could create artificially radioactive isotopes by bombarding virtually any element with sufficient numbers of neutrons. He and his small group arranged to capture neutrons in the nuclei of atoms. The nucleus would absorb the extra neutron, creating an isotope of that element. He also found that the slower the neutrons flew at the nuclei, the more effective they would be and that the rate of absorption was different for each element. What was to be particularly important was what happened when they did that to ordinary uranium.

Fission, the force behind the atomic bomb, was discovered in 1938. In December of that year, weeks after Fermi left Rome for the Nobel Prize in Stockholm, two physicists, Otto Hahn and Fritz Strassmann, working on discoveries by Fermi, established that if you blast a neutron into uranium you got a lighter element, barium, and the release of a great deal of energy. They could not explain what was happening, but two other German physicists, Lise Meitner and her nephew Otto Frisch, Jewish refugees from Nazi Germany (he living in exile in London, she in Stockholm), explained it a month later: they were splitting the atom, and $E=mc^2$ and all that. Frisch called it "fission," after a term in biology.

That autumn, Fermi had been awarded the Nobel Prize for his work, took his wife and children to Stockholm for the award ceremony, and kept going. While Fermi wasn't Jewish, his wife Laura was, which made his children also Jewish. The Nazis had already imposed a law in Germany declaring that if you even had one Jewish grandparent or spouse, you were a Jew and your fate was at best uncertain, threatening the entire family. Fermi had job offers from several American universities, and he took the one at Columbia. From America, Laura tried to get the rest of her family out of fascist Italy, including her father, a retired admiral in the Italian navy. He was sure that as a hero of World War I that he would not be harmed. He was murdered in Auschwitz.

In 1940, Frisch and Rudolf Peierls outlined how the knowledge of fission could be turned into a weapon. Arthur Compton, president of the University of Chicago, who also saw the potential in the work of Frisch and Meitner, and fearing the Germans did as well, began bringing physicists from all over the world to the university, including Fermi, Eugene Wigner, Edward Teller, Harold Urey, Leo Szilard, and Glenn Seaborg.

Moving from Columbia to Chicago in 1942, Fermi took over the squash court under the west stands of Stagg Field, the abandoned football stadium—the school had given up varsity football in 1929—and built a huge sphere twenty-six feet in diameter. It consisted of 400 tons of graphite and six tons of uranium metal and fifty tons of uranium oxide. When the control rods were withdrawn, Fermi started the world's first chain reaction. This also proved that it would be possible to build a city-destroying bomb.

Work continued through the war at Los Alamos, and even after it ended with the bombing of Hiroshima and Nagasaki, Fermi continued to work at Los Alamos through the summers.

Having decided on who he wanted for an advisor, Garwin went to see Fermi.

"I explained to Fermi that I had a lot of experience with electronics (vacuum tubes), and that I was also good at mechanical drawing, and that I could operate a lathe, milling machines, a drill press, and the other tools that were useful in making physical apparatus, and would he be willing to have me work for a while in his lab to see whether I could be of use to him," Garwin wrote. "Fermi was also a gambler and acquiesced. He took me into his lab—room 286—Ryerson Physics Laboratory, where important work was done during the Manhattan Project."[4]

Physics can sometimes look like high-class plumbing. Physicists need elaborate equipment, often custom-made, to do experiments. Garwin said Chicago was particularly fortunate because most of the technicians and mechanics had either stayed through the war or, if they went to work elsewhere on the Manhattan Project, returned. They included glassblowers, machinists, and technicians. When things were going well, a physicist could order up a piece of equipment and someone would make it either to the physicist's specifications or into workshop designs. Garwin's offer would have been hard to turn down.

Fermi was not alone in the lab. One collaborator, Leona Woods Marshall, who worked with Fermi at the Stagg Field experiment and then went with him to Los Alamos, returned to the lab. They were working on positronium—an atom consisting of a positron and an electron, orbiting each other. The positron is the electron's antimatter particle. Another collaborator was Jack Steinberger, who was off in a corner doing something with cosmic ray particles: he would eventually receive a Nobel Prize.

"Fermi showed me what seemed to me to be heaven," Garwin wrote, "a lathe that he had used himself (along with a young technician), racks of equipment, and benches where one could do *experiments*." There were two tiers of wooden benches and blackboards since the room was also used for lectures.[5]

"Professor Fermi," Garwin told him, "I can do this. There is no reason for you to take your time. Why don't you think about other things?" The first thing he did was invent a better way for Fermi and Marshall to keep track of the electrons their experiments produced, using a coincidence circuit. "I'm good at that," he said.[6]

For his own dissertation, Garwin built other counters, making use of mothballs (naphthalene). He was researching gamma rays and the properties of radioactive nuclei using sodium, as in sodium chloride, salt. He created a Rube Goldberg device to measure the angle of emission of gamma rays following the decay of a radioactive substance.

To create the radioactive elements he would have to go to the Argonne National Laboratory outside Chicago regularly, which he did not like to do, so he adopted a way of working around the problem by visiting Argonne only occasionally, transporting the slightly radioactive material back to Chicago in the trunk of his car, and working on it in the lab. He used an eyedropper to get the sodium on a very thin lacquer film on an aluminum ring. It was so thin that one had to tilt it against the light to detect its presence. The purpose was to detect the electrons as the sodium decayed and transformed into magnesium, and to match that with gamma radiation. The salt provided the sodium, and the mothballs produced naphthalene crystals that were used as a detector glued into aluminum tubes. He produced gamma ray measurements less than a hundred millionth of a second, some 1,000 times faster than any other device.

He would irradiate the sodium, put it in the back of the car, and drive back to the lab about thirty minutes away, the material placed in a Lucite carryall he designed and built. The technique he invented for measuring the decay became standard in physics for twenty years. The calculations were done on Marchant mechanical calculators. Lois would put the numbers on spreadsheets at night so the data could be analyzed the next morning. The tools he created would be used later in his most famous experiment.

Garwin was awarded his PhD in 1949 after only two years, and despite a school policy against hiring its own former graduate students, the university gave him a job.

Because the cyclotron at Chicago worked twenty-four hours a day, Garwin worked odd hours. Sometimes he would come home for dinner and then, after a few hours, go back to the laboratory. It was within walking distance, but he would come home late.[7]

Fermi asked him to help with his work on a new model for the nuclei of atoms, later called the shell model. He sketched out what he wanted Garwin to do. In a rare moment of introspection, Garwin didn't think he could "pull it off,"[8] and when Fermi came back to see him a week later and found Garwin hadn't started, he gave the project to Maria Goeppert Mayer, a theoretical physicist volunteering in his lab because her husband worked at the university and Chicago also had an anti-nepotism policy. She later shared a Nobel Prize for the work.

"Opportunity missed," Garwin reported in his unpublished autobiography.[9]

In the meantime, the chancellor of the university, Robert Hutchins, had brought in some scientists who had worked on the Manhattan Project, and the connection between the university and Los Alamos was strong. Fermi spent summers there.

One day in Chicago, Lois recalled, Garwin asked Fermi a question. A strange look came over Fermi's face, and he told Garwin he couldn't talk about it. Later, he told Garwin, he could answer the question in Los Alamos, not Chicago, and if Garwin really wanted to know the answer he should come to Los Alamos in the summer.

He would get Garwin a security clearance, he said. And, Los Alamos paid. Garwin and his young family moved to New Mexico for the summer of 1950.

"Jeffrey was born in November of '49, so we took him along," Lois said. "We traveled by car from Chicago to Los Alamos. It was before the days of [child] car seats, so what we did was, we filled the space between the back and the front seats level to the back seat with boxes and suitcases and whatever. Then we took the mattress from Jeffrey's crib, and we put it on top. Then Jeffrey had a playpen. And it worked very well."[10]

They loved New Mexico.

The road to Los Alamos, situated atop a remote mesa, hugged the side of the cliff. Originally, it was unpaved with few safety fences on the side, but by the time the Garwins got there it had at least been paved and was less scary than it was when the Manhattan Project community arrived years earlier. The Garwins moved into a government-issue, two-bedroom apartment they rented from Los Alamos's management company, Zia, equipped with the same Southwestern style as every other apartment at Los Alamos with the same furniture.

Garwin continued to consult for years at Los Alamos, and Jeffrey fondly remembers later trips. He spent around ten summers at the lab.

"There is nothing better than Los Alamos for a kid in the 1950s. I have very fond memories of that," Jeffrey said. "So [Garwin] would work at the lab during the day and I hadn't the foggiest idea about what was going on except my earliest memory of Los Alamos is driving up the hill and sleeping in the back of the car and being awakened when we came to the checkpoint where there were these spotlights and guys with machine guns. . . . I got to spend my days exploring the canyons and playing with friends and riding bikes and stuff like that. My dad would come home fairly early from the lab and he had lots of friends in Los Alamos and they would all come over and some of them were pretty famous people. I saw . . . I'm sure I saw a couple Nobel Prize winners in Los Alamos and lots of other people who were just plain famous. Of course I didn't know them then, I just categorized them as whether I liked them or not." One of Garwin's closest friends was Harold Agnew, who eventually became director of the Los Alamos lab.[11]

Lois would drive Garwin to work in the morning with a kid or two in the car and pick him up in the afternoon. No one in the family was allowed in the building where he worked. There were picnics. Jeffrey remembers Teller's notable eyebrows and that one day his father gleefully demonstrated how a rolling toy worked to Fermi. The toy was shaped like a drum, and if you rolled it away from you it would roll back. Physicists loved to play with it, Jeffrey said.[12] Not physicists' children. Physicists.

Garwin couldn't resist giving geology lessons, how formations came about. He would look at people's houses for cracks in the foundations and try to determine how to fix them, Jeffrey remembered.[13]

One of his first projects at Los Alamos was the theory of "fratricide," the potential of emasculating one nuclear explosion by blowing up another nuclear explosion nearby. It turned out not to be a dingbat idea. Later, Fermi showed him how to do the calculations.

Garwin spent much of the rest of the time learning what it was like living in one of the most security-conscious places on the planet, and the first week in the guarded library reading up on what had happened in the years since 1943 and after Hiroshima and Nagasaki in 1945.

"So I was quickly *au courrant* in the world of fission weapons, and I read some of the considerations and work that had been done on the possibility and the potential design of thermonuclear weapons," he wrote.[14]

He and Fermi shared an office in the theoretical division, about the size of a walk-in closet. It had two desks facing each other and a safe for classified materials. They shared the combination. They kept a notebook for secret documents—or rather Garwin did for the both of them. Fermi didn't want to bother.

Lois knew that Los Alamos was involved in nuclear weapons but never discussed Garwin's work with him there. He was very careful. "Let's put it that way," Lois said. "I really don't know that I discussed it with Dick, but certainly at parties at Los Alamos, they talked about it."[15]

It was the beginning of the Cold War. Europe was divided. That spring, the Soviet Union exploded its first atomic bomb, a feat largely made possible by Soviet spies at Los Alamos. Garwin said the state of the world did not directly affect him there, but several months before he arrived in New Mexico, President Harry S. Truman had announced that the United States was proceeding with development of nuclear weapons "including the so-called hydrogen or super-bomb."[16] On June 25, 1950, the Korean War broke out.

Teller, almost frantic with anticipation and frustration, decided to drop in to see what insights he could get from Fermi. Garwin was in the office.

CHAPTER THREE

THE SUPER

Everyone had an idea of what a hydrogen or thermonuclear bomb would look like even if no one was sure it would work. It would be either a cylinder or a sphere of steel with thermonuclear fuel and isotopes of hydrogen—supercooled liquid deuterium or deuterium mixed with tritium, or tritium, a radioactive isotope of lithium—inside. A nuclear bomb would explode and somehow raise the temperature of the fuel to tens of millions of degrees, producing pressure, causing the nuclei of the fuel to fuse and release unspeakable amounts of energy. If it failed to get hot enough, all the energy would be wasted and the device would fizzle.[1]

Fusion was known seven years before the discovery of fission. Hans Bethe eventually won a Nobel Prize for describing it. Two light atomic nuclei, say hydrogen, the most common element in the universe, fuse under enormous force—for instance gravity—and turn into helium, giving off huge amounts of energy. It's the furnace of the sun and the stars. But getting that to happen on demand on Earth was a huge problem. It had never been done before, and many thought it could not be done. Fusion in the sun is produced by the force of gravity created by its mass over billions of years. How do you reproduce a force that massive on little Earth?

Like fission, a fusion bomb doesn't work as an explosion unless you get a huge number of atoms to release energy instantly. The preferred fuel—tritium—needed to be made in a nuclear reactor. Half of it would decay from radioactivity in twelve years, and each gram of tritium made would take the place of eighty grams of plutonium needed to build nuclear weapons like the one that destroyed Nagasaki in 1945. Then, how do you light the fuel? There were a substantial number of physicists who believed it couldn't be done.

There were a substantial number of physicists who believed it *shouldn't* be done. They had a history.

World War I was not only the first mechanized war, but also a war of chemistry. The German Nobel laureate Fitz Haber learned how to synthesize ammonia from nitrogen, which produced fertilizers that have fed the world since. He also produced explosives and invented poison gas for Germany. The Allies enlisted their own chemists and retaliated with gas of their own, and millions of soldiers died horrible deaths.

World War II was the physicists' war. It came in the midst of perhaps the most productive and creative time in the history of science, a golden age of physics. Einstein had moved the world from Isaac Newton. Werner Heisenberg and Niels Bohr, among others, moved the world beyond Einstein to quantum physics.[2]

It was Einstein who recommended to President Roosevelt an emergency project to build the atomic bomb. There was no formal relationship between the scientific community and the government at the time, so Einstein had to have a letter delivered to the president by one of Roosevelt's friends. In the letter, he warned that the Germans were undoubtedly aware of the advances in physics and had the scientists (Werner Heisenberg, creator of the Uncertainty Principle, a key to quantum mechanics, among others) to do it. The letter was written in July 1939 by Leo Szilard and Alexander Sachs, signed by Einstein and delivered by Sachs.[3]

Roosevelt thought the atomic bomb was worth exploring and turned to the government's bureaucracy, starting with the National Bureau of Standards and Lyman Briggs, its chairman. The idea that the United States had to act on the weapons went through various committees, including one called the Advisory Committee on Uranium, which eventually dropped the "uranium" from the name to hide its purpose.

It finally found its way three years later to Vannevar Bush, the most influential scientist-politician in the country, and Bush turned it over to the US Army Corps of Engineers' Manhattan District, hence the name Manhattan Project. They assigned control to Colonel—later

Lieutenant General—Leslie Groves. To this day, Groves is remembered by scientists at Los Alamos as a "fat buffoon,"[4] but a case could be made that he was one of the most competent military officers in American history. Groves appointed J. Robert Oppenheimer, one of the most extraordinary personalities in science, to be the science leader. Most physicists admired Oppenheimer as a scientist, a scholar, and as a person—dashing, handsome, cultured, with a remarkable memory and an ability to remember names and people—but few thought he was capable of leading such a large team in such a great quest. He had never done anything like it before, and no one, except Groves, thought he had it in him. They were wrong. Oppenheimer would gather many of the world's greatest physical scientists to work for the government under trying circumstances. What he built was unprecedented in history.

Soon, extraordinary things began to happen.

One day in 1942, physicists and graduate students started to disappear from labs all over America. A famous scientist or some promising graduate student would be there on Thursday but gone on Friday. Almost no one knew where he (or she) went off to or why. Investigation would show that in many cases, their families had eventually disappeared with them. Those recruited were not told where they were going or what they would do when they got there, only, as one said, that Uncle Sam wanted them. Secretly, they were heading to places like Los Alamos, New Mexico (called Site Y in official documents), Oak Ridge, Tennessee, and Hanford, Washington, where Groves had, in a matter of months, created three towns and two huge processing plants.[5]

Many of those recruited, including some from the University of Chicago, stayed where they were but moved their offices or spent a lot of time at places like Enrico Fermi's new facility—cover name Metallurgical Laboratory. Some would just be out of town for long periods of time. New labs were built, and old ones acquired safes, guards, and locked doors. The University of California at Berkeley was courted to participate. A few scientists even acquired aliases to hide their identities. Fermi was known as Henry Farmer.[6] He also acquired a bodyguard.

The plutonium production plant in Hanford grew from zero to 40,000 workers in eighteen months, becoming the fourth or fifth biggest city in Washington State. Virtually none of the people who worked there knew what it was they were doing, but men like Fermi and Eugene Wigner had offices there.

Hundreds, later thousands, were ordered to report to 109 East Palace Avenue, near the plaza in Santa Fe, New Mexico, a ground floor adobe office in what was once a Spanish fortress. There, they were processed by Dorothy McKibbin, a local widow hired by Oppenheimer, who eventually would run the project as much as he and Groves did despite not knowing until the end what everyone was there for. They filled out forms, were given security briefings, and then taken out the back door so they could not be seen to busses or cars that would drive them to Site Y, called by everyone "the Hill,"[7] which sat atop the mesa northwest of Los Alamos. It was a site Oppenheimer knew well from childhood because he and his brother Frank had gone camping and horseback riding there, and he had recommended it to Groves because of its remoteness. A boys' school was at the site. Groves bought it. The transport, occasionally a convoy, would drive through the high desert, beige hills sprinkled with green dots of piñions and junipers, increasing in altitude, past the reservations and pueblos of the Tesuque, Pojoaque, San Ildefonso, and Santa Clara people. Then up a winding dirt road that clung to the side of a mesa, surrounded by purple-brown buttes and deep canyon walls, the Jemez Mountains lurking to the west. The last miles of the ride were beautiful and frightening.

The early ones arrived in the middle of a dusty—sometimes muddy—construction site with few amenities, insufficient housing, and disorganization. Life would often be difficult, and while it was something of a camp for some of the world's most famous scientists, it sometimes suffered from the unique problem of too many geniuses in a closed space. Things did not always work smoothly or happily. But the job got done. It would be one of the largest collections of scientific geniuses in history. Certainly almost all the great giants of physics were either based there, had consulted there, or had visited there, the greatest collaboration between science and government ever.

Throughout the project, people kept pouring onto the Hill, including eventually local construction workers, soldiers with varying skills, doctors, teachers, technicians, even housekeepers. It became a town.

Technically, Los Alamos, just one part of the Manhattan District, was an army post, with Groves in charge. Occasionally, someone in the army came to New Mexico to try to make it work like an army post, for instance requiring exhausted, rebellious civilians to march in review, but mostly they failed.

Groves never pushed the issue. He worked almost exclusively from New York, leaving Oppenheimer in charge. Oppenheimer was on the phone regularly to him, and the cooperation between the two men was close, respectful, and efficient.

Science had gone to work for the government as an act of patriotism. Oppenheimer had little trouble recruiting scientists. He remembered:

> Almost everyone knew if it were completed successfully and rapidly enough, it might determine the outcome of the war. Almost every knew that it was an unparalleled opportunity to bring to bear the basic knowledge and art of science for the benefit of his country. Almost everyone knew that this job, if it were achieved, would be a part of history. The sense of excitement, of devotion, and of patriotism in the end prevailed. Most of those with whom I talked, came to Los Alamos.[8]

Scientists from around the world arrived as well, many from Britain, which for a while was ahead of the United States in nuclear research. Even Niels Bohr and his son Aage (both would win Nobels) arrived from Sweden, where they had taken sanctuary—Bohr's father was Jewish, and their native Denmark was occupied by the Nazis.

They came for a few weeks under heavy guard. It would have been no trouble arranging several poker games with only Nobel laureates, present or future, playing.

People working on the project never talked about their work. They couldn't speak to anyone outside the labs about it—even to their spouses. Mail was censored, phone calls monitored, visitors discouraged. They put up with it. Herbert York, a Native American graduate student who worked at Berkeley, remembered that even the word "uranium" was never spoken out loud even among themselves. It was called "tube alloy." The same secrecy prevailed at Columbia University in New York and the Radiation Laboratory at MIT.[9]

On July 16, 1945, their work was tested at Alamogordo, New Mexico, the so-called Trinity test of a plutonium bomb. Physicist Richard Feynman sat on the hood of a Jeep, watching the mushroom cloud erupt into the sky, while he banged on bongo drums in an explosion of emotion.[10] Isidor Rabi took out a bottle of whiskey from the trunk of his car, drank deeply, and passed the bottle round. Oppenheimer famously quoted from the *Bhagavad-Gita*, "Now I am become Death, the destroyer of worlds."[11]

In the physics community, the Manhattan Project was not controversial until after it succeeded.

The minute the Trinity bomb exploded, their product became a weapon, and weapons were under the control of the military. It was no longer science; it was war bordering on genocide, and physicists— along with chemists, mathematicians, and engineers—were largely responsible.

Many report that the days after Trinity, the most prevalent emotion was depression, partly because they had worked so hard and now it was over, a predictable letdown, and partly because of what they had done.

After the blast, Feynman saw his friend Robert Wilson sitting slumped. "Why are you moping about," Feynman asked.[12]

"It's a terrible thing we have made," Wilson replied.

Oppenheimer had sat in at meetings where the military selected potential targets. He knew Hiroshima and at least one other city, and perhaps eventually even flammable Tokyo, would soon be ashes and hundreds of thousands of Japanese, maybe even a million, would soon be dead, blown apart, fried, melted, vaporized, or zapped by radiation.[13] He also knew that the alternative to dropping those bombs would have been an invasion of the Japanese home islands, which would likely result in the death of a million people, many of them American servicemen and untold numbers of women and children. The army ordered half a million Purple Hearts.

A month before Trinity, a group of physicists at Chicago urged the government to demonstrate the power of the bomb to the Japanese and give them a chance to surrender. It was rejected.[14] After Hiroshima and Nagasaki, the first attempt to control the weapons was organized. When Congress was considering creating the Atomic Energy Commission (AEC) in the aftermath, Manhattan Project scientists lobbied against it because the bill would put the weapons completely in the hands of the military. They succeeded, and the AEC was eventually put under civilian control, although in the end it is debatable if that made a difference.

The day after the war ended, Oppenheimer came to Edward Teller's office in Los Alamos and said that with the war over, "there is no reason to work on the hydrogen bomb."[15] Teller, who had been working on it full-time for two months, was stunned. He later wrote that he thought Oppenheimer had become unhinged, and pointed to the *Bhagavad Gita* quote as an example. Teller took Oppenheimer's statement as notice that he had essentially been fired. Opinion at Los Alamos also had turned swiftly against the hydrogen bomb almost from the moment Hiroshima was destroyed. Teller left for Chicago.

In a speech Oppenheimer gave when the laboratory was awarded a medal for its wartime service, he warned that if bombs became part of the world's arsenal, "mankind will curse the names of Los Alamos and Hiroshima." Many who worked on the bomb agreed.[16]

The scientists who knew the most about the weapons wanted an international effort to control them, to put the genie back in the bottle, as the cliché goes, and began organizing to do just that. They formed the Federation of Atomic Scientists (later the Federation of American Scientists).

For many of the scientists who quickly headed back to their abandoned offices and students, the making of the bomb was the greatest, most important time of their lives, which probably changed them all. Most stopped doing weapons work, with the notable exception of Teller. There was too much good physics to do elsewhere. The generation of scientists who worked on the Manhattan Project would dominate American science and wield enormous power in Washington for a generation.

But they were scarred.

"In some sort of crude sense," Oppenheimer wrote, "which no vulgarity, no humor, no overstatement can quite extinguish, the physicists have known sin; and this a knowledge they cannot lose."[17]

On July 12, 1947, Los Alamos was transferred to civilian control. When the war ended, the United States had exactly one atomic bomb left.[18]

Whether or not to build a hydrogen bomb was a great moral issue that roiled the physics community as the war ended and echoed through history ever since. Most of the physicists at Los Alamos and the other stations of the project went back to whence they came, universities and industries. To get a thermonuclear—the technical term—weapon built would involve enticing some of them back, and that proved difficult. While some supported building the weapon, many vehemently objected on moral grounds. Most, perhaps, were undecided.

There were few moral objections to building the atomic bomb while America was at war for survival against an enemy with the talent and knowledge to attempt to build one itself. If they knew such a weapon was possible, so surely did Germany and certainly German physicists like Heisenberg, who worked under the Nazi regime, knew it. Physicists were entirely on new ground.

Teller, a refugee from Hungary who had relatives trapped in

the Holocaust, thought Stalin and the Soviet Union were genuinely depraved. He was honestly and deeply afraid of their intentions, was sure that if America didn't build what Teller called the "Super,"[19] the Soviets would. He even told a friend that if the bomb was not built, he envisioned himself a prisoner of war in a Soviet prison somewhere. He was not alone; many of the brain power that built the atomic bomb were refugees; a large number were Jewish with personal horror tales of their own, men who knew firsthand the dangers a psychopathic dictator like Stalin presented. Their fear was real and could not easily be dismissed.

If the Russians built the Super first, America's situation would be "hopeless," Teller said.[20] They could easily destroy the US. The weakness in the argument, which Teller seemed to ignore, was that the US eventually would have a large stockpile of atomic bombs in various iterations that could quickly destroy the Soviet Union in retaliation. The US did not need a hydrogen bomb to obliterate Russia—and the Russians knew it.

"That's the one monomaniac I know with more than one mania," Fermi once cracked to Garwin after a visit from Teller.[21]

Teller was supported by Nobel laureates Luis Alvarez and Ernest Lawrence at Berkeley who, perhaps less frightened, still believed strongly that the defense of the nation depended on America building a hydrogen bomb first.

Opposition to the H-bomb came from men like James Conant, the president of Harvard, who had worked on the Manhattan Project, and David Lilienthal, chairman of the Atomic Energy Commission. Lilienthal was unimpressed with the argument that the US had no choice but to build the bomb. He thought rather that perhaps the problem was the US maybe just wasn't smart enough to see an alternative.[22] He likened it to the Maginot Line that failed the French in World War II.

The military was almost entirely in favor of the bomb, naturally, although General Omar Bradley was doubtful it was a good idea.

The most important of the scientists involved in this weapons work was Oppenheimer, who early on was skeptical but had not made up his mind entirely about whether to build one. His hesitancy would lead to one of American science's great dramas, the testimony of Edward Teller that would cost Oppenheimer's security clearance and Teller his reputation.

Fermi had originally opposed development of a fusion bomb and refused to be involved. He would later change his mind.

The indictment of Klaus Fuchs, a British scientist who worked on the atomic bomb at Los Alamos as a Soviet spy and who helped the Soviets catch up in the nuclear race, made life difficult for the opponents.

Teller kept pushing for a major project to build the Super, but Norris Bradbury, the new director at Los Alamos, did not agree.

In those days, the AEC produced nuclear weapons for the military, so to get a handle on the Super, the AEC asked its high-level General Advisory Committee (GAC) to investigate the possibilities. Oppenheimer was the chair, and members included Isidor Rabi and Fermi. They met in Princeton on October 28–30, 1949.[23] The question they were posed was, should the US build a hydrogen bomb? The answer from the GAC was an emphatic *no*! Fermi and Rabi added to the majority report a minority report that went even further. They called it intrinsically evil, a weapon of genocide.

"The fact that no limits exist to the destructiveness of this weapon makes its very existence and the knowledge of its construction a danger to humanity as a whole. It is necessarily an evil thing considered in any light," they wrote.[24]

"Necessarily, such a weapon goes far beyond any military objective and enters the ring of very great natural catastrophes. It is clear that the use of such a weapon cannot be justified on any ethical ground which gives a human being a certain individuality and dignity even if he happens to be a resident of an enemy country."[25] Scientists rarely use the word "evil," a word with no scientific context. Nonetheless, evil is what they saw.

The reference to limits refers to the fact that in theory, one could build a fusion weapon of unlimited size, one say large enough to destroy the world. You would create a new sun. The only limit was the amount of fuel available.

Oddly, the main report came out against the bomb for technical, not moral, reasons. The committee did not see how it would be possible to build such a thing. If so, that finessed the moral issue. They could dismiss the weapon without going into the swamp of morality.[26]

No target, Oppenheimer said, was big enough to justify its use.

Despite the report and probably influenced by the military and for political reasons, President Harry S. Truman gave the go-ahead. It is likely he did not understand the ramifications of what he was doing, but the order went out anyway.

"A lot of the push for going forward with the bomb was in, my view, almost ludicrous, because at the time, when Truman said we're going to make a crash effort in building a hydrogen bomb, we didn't have the faintest goddamn idea of how to do it," physicist Marvin "Murph" Goldberger said. "The original Teller concept had been disproved. We announced this as a program, but there, in a certain sense, was no program."[27]

Nevertheless some of the scientists who opposed building the bomb agreed to go back to work, including eventually Bethe. Earlier, Bethe was so adamant about having nothing to do with the bomb that he would leave a meeting if the subject came up and would not enter an office unless he was assured it would not enter the conversation.[28] He eventually changed his mind, apparently also afraid of Soviet advances, and was assigned to do the theoretical work just as he had on the fission bomb.

But that still left a huge issue, whether the hydrogen bomb could be built at all, whether the laws of nature permitted triggering the reactions necessary to produce a thermonuclear explosion outside of a gravitational field from a gigantic body like a star. One thing was clear: the project would require calculations well beyond anything science or engineering had yet produced. The early mathematics were performed in part by ENIAC, the world's first programmable electronic computer at the University of Pennsylvania. In 1945, using a million IBM punch cards, Los Alamos mathematicians tried to test the original theory of the hydrogen bomb on ENIAC at Penn's Moore School of Electrical Engineering. It worked to an extent, but ENIAC wasn't powerful enough to handle the numbers required to actually design a fusion device.

The mathematician John von Neumann was working on a successor to ENIAC at the Institute for Advanced Study in Princeton called JOHNNIAC, but it wasn't ready to handle the computing requirements for the Super. Another, MANIAC, was under construction at Los Alamos. Carson Mark, director of the theoretical division at Los Alamos, brought Nicolas Metropolis from Chicago to run the computing operation. Metropolis had led the ENIAC calculations in Philadelphia.

The lack of computing power was one of Teller's problems.

Until the electronic computers were ready, most physicists used desktop calculators, mostly made by Marchant—devices with buttons and a hand crank, although later some were electric-driven. The word "computers" was then defined as mathematically gifted humans (usually women) who did the calculations by hand on the Marchants. At Chicago, whenever Garwin needed calculations for processing the data for his thesis, Lois Garwin was his computer.

"After I had worked out the arithmetic procedure (what would now be called a 'program') . . . Lois, would come in the late evening and take the numbers from the laboratory notebook, write them on an accounting spreadsheet, put them two at a time into the Marchant calculator, writing down each intermediate result," Garwin said.[29] He was happy to let her do it because, he said, she was at least as good as he was in entering the numbers on the keyboard and he could work on the physics. She would have the results back in the morning. The arrangement was the same at Los Alamos.

The computer working on a fusion bomb at Los Alamo with Fermi was Miriam Planck Caldwell. Fermi had done some calculations specifically aimed at the potential fizzle. Results were discouraging. A major problem requiring calculation was the diameter of the cylinder containing the fuel, Garwin recalled. If it was too small, energy would escape and not produce the fused nuclei they needed, the "thermonuclear burn." Too large a diameter and the reaction would be quenched by radiation. Teller and his colleague, Stanislaw Ulam, had no idea what that diameter was and how to test it. They couldn't do the math.

Without that calculation, they could not run a test. If they tried and the test failed they had no way of knowing whether the design was wrong, the test was wrong, or a thermonuclear device was instead impossible. They would not know why the test failed. Teller had no doubts it was possible. But as time went on, it seemed less feasible. Every time Teller made a report, they took a step backward. Even Teller admitted he was stumped and, in 1947, suggested they wait to do extensive work on the bomb until they had the computing power.

Teller had been assigned to work with Stanislaw Ulam by Oppenheimer. They did not get along. Ulam didn't know how to build a Super either, and every morning, he would throw one die of a pair of dice to generate a random number for the equations, guessing coefficients for the fate of neutrons in the reaction. They would form the numbers for the motion of particles that Caldwell and the other "computers" took home at night. Teller, appalled that the calculations were essentially based on a crapshoot, loudly recommended the results be ignored, in part because they also looked like they would produce thermonuclear fizzles.

Fermi tried to help. As Garwin described it:

Fermi would use an accountant's spreadsheet, filling in the top row with various initial conditions and carrying out the calculation by using a slide rule and a desk-top mechanical calculator (a Marchant, as I recall) to advance each time step. After seeing that the calculation progressed properly, Fermi and Ulam would call in [Caldwell] who would work overnight on the problem, bringing it back in the morning. Fermi and Ulam would graph the results, decide to change some input condition or assumed cross-section or other parameter, and go on to other things, while Mrs. Caldwell worked on the next day's calculation. The results were discouraging.[30]

Edward Teller was part of what physicists sometimes called the "Martian invasion"—a quintet of startlingly brilliant men who all came from Budapest (although, the joke went, probably from another planet). They included von Neumann, Leo Szilard, Eugene Wigner, and Theodore von Karman, an engineer who was working in fluid dynamics and who would later help rocket men to the moon. They went to the same high school and lived in the same neighborhood. They spoke a language, Hungarian, which no one else spoke and which was not based on Latin. They were all Jewish. These men had thrown themselves into building the atomic bomb lest the Germans get to the bomb first.

They assumed—correctly as it turned out—that Heisenberg was working for Hitler and probably on the bomb, although in fact, Heisenberg never came close to building a bomb. And the German nuclear weapon project, called the Uranium Club, was actually led by a physicist, Kurt Deibner, who was no Oppenheimer. They were also hampered by the destruction of a heavy water plant by Norwegian commandos. Heavy water is water with a high concentration of the hydrogen isotope deuterium. Heisenberg would later claim he had deliberately stalled the project to make sure Hitler did not get the bomb before the war ended. Garwin believed he simply made a crucial calculation error early on in his research and never knew it.[31] Heisenberg claimed to be astonished by Hiroshima. Strassmann and Hahn also were still in Germany. Hahn worked on the German project, and Strassmann never did.

Teller had been finally sidelined so he could work on his idea without getting into anyone's way.

Ulam was almost the opposite. A Polish Jew who also had fled Hitler, he was a mathematician where Teller was a physicist. Ulam was, Kenneth Ford wrote, laid back, where Teller was intense. Ulam had a sense of humor; Teller had none. They were an odd marriage. Teller accused Ulam of having a dozen ideas a day—almost all crazy. Ulam had no idea when he came up with a good one, Teller said, and it was Teller's task to find them and exploit them.[32] Oddly, the same thing was said about Teller. He had, others at Los Alamos said, hundreds of ideas a month and a couple of them were brilliant, the others not so much.

"Now Ulam, like Teller, was a person who was always interested in something new and novel, but he didn't care to see them worked out. He just wanted to go on to his next idea, whatever it was, whereas Teller wanted to see that everything that he proposed was actually built," Garwin said.[33]

They detested each other.

There were several theoretical ways to build a bomb—if such a bomb could be built. They each had nicknames. The point was to start a reaction that would cause thermonuclear fuel, liquid deuterium probably, that would lead to an explosion rather than just have the whole complex, expensive device sputter out. The device also had to be useful: it had to fit in a bomber, resist exploding from jostling and hard landings, and blow up on cue. It was generally conceded the only way to start the process was to use an atomic—fission—bomb as the igniter, like throwing a lit match into a can of gasoline. One way was to put the thermonuclear fuel in the middle of the fission bomb. There was the Alarm Clock or Layer Cake model, which had alternating layers of fission and thermonuclear fuel, an idea model eventually used by the Russians. There also was the inappropriately named Yule Log, a cylinder of thermonuclear fuel. No one knew which was best.

One day after New Year's 1951, Ulam had an idea. His wife remembered him sitting, staring out the window with a strange look on his face. She asked him what was wrong, and he told her he had figured out how to make the Super work. She was appalled. She did not want the Super to work.[34]

The thermonuclear fuel, it was conceded, would be hard to light and in short supply. The solution to that problem, Ulam concluded, was to use compression—squeezing the fuel so you'd need less of it.

After working out some of the details, Ulam took his idea to Norris Bradbury, the director of the lab. Bradbury brushed him off.

Unable to procrastinate any longer, on March 1, 1951, Ulam met with Teller for two painful hours to convince him that his plan would work.[35]

"'Look, Edward, I have a good idea,'" Garwin quoted Ulam. "'Let's use the energy from a fission explosion to compress material.' He was going to use the shock from a nuclear explosion to do this. And he would use hydrodynamic lenses, which are what they used in the Trinity-Nagasaki weapon. Those lenses were constructed of a combination of fast and slow explosives that contain and drive the energy from a blast in a designated direction."[36]

Teller earlier had rejected compression. He had thought about it for more than ten years, starting with a walk he had with Fermi in New Jersey in 1939.[37] His idea of a Super did not make use of compressed fuel. "Compression makes no difference," he said.

"Edward was a fine theoretical physicist, very ingenious ... you know, when you could pin him down to one idea at a time, he could really do a lot of working it out," Garwin said.[38] Teller may have been inclined to reject compression, in part, because the idea came from Ulam. "You can waste a lot of time talking to Stan," Teller said. But Teller was in despair; nothing else was working.

Ulam initially thought the pressure from the blast of the nuclear device, if aimed properly, would be enough to ignite the fuel. Teller told Ulam it was a bad idea. "Most scientists are delighted to show that somebody else has had a bad idea. I can imagine his enthusiasm," Garwin said.[39]

Teller, however, thought he had a better way to do it, something he once knew and had forgotten—use X-rays from the blast for the compression. At the temperatures of a fission bomb, 80 percent or more of the energy exists as soft X-rays. As physicist Kenneth Ford points out, very hot radiation acts like a "substance," a physical entity that can flow and push things, like a powerful piston. Another advantage was the X-rays would move a lot faster than the pressure wave. They can be channeled by the lenses. A powerful enough blast surrounding a container of thermonuclear fuel would implode it and possibly trigger the weapon.[40] That was a two-stage bomb, fission-fusion. To make an even larger bomb, a third stage, a fission bomb, could be added, fission-fusion-fission.

Teller realized it could work. Now he had to figure out how.

"Hans Bethe says that it's a kind of miracle, this radiation implosion, this idea," Garwin said.[41] In hindsight, it seemed so obvious. Garwin guessed that it would have taken another year or two without Ulam's satori. To Teller's credit, he was willing to change his mind even if the idea did come from Ulam. In the Soviet Union, Andre Sakharov and his team of bomb designers had about simultaneously come to the same conclusions, as had scientists in France, Britain, and later China.

Eight days later, Teller and Ulam wrote a paper explaining their theory and suggested two ways to make it work. The title of the paper was "On Heterocatalytic Detonation I. Hydrodynamic Lenses and Radiation Mirrors."[42]

"The scheme," the paper said, "then depends concentrating, as much as possible, the energy released by the explosion of a fission bomb in the mass of the principle assembly and *doing so as to achieve a high compression* in this mass." [Original italics.]

Copies of the paper, with most of it redacted, have since been published. Oppenheimer, previously a skeptic, called their idea "technically sweet."[43]

The two men spent much of the rest of their lives debating who did what and who should get credit. Teller would win the public relations war.

Richard Garwin, in Chicago and later in Asia, did not know of the breakthrough when he arrived in Los Alamos that summer.

CHAPTER FOUR

GARWIN'S DESIGN

According to legend, one day Enrico Fermi walked into the Fuller Lodge at Los Alamos and told the Manhattan Project scientists at the table he wanted to tell them about the first "real genius" he had ever met. According to the legend, several of the men at the table thought Fermi was going to single them out and a few chests expanded in anticipation.

"Dick Garwin," Fermi said.[1]

Garwin already had established himself as an unusually bright student, sure of himself and very quick. When Garwin asked Teller in Los Alamos in May 1951 if he could help, Teller jumped immediately to take advantage of his skills, and pointed Garwin to the documentation he needed to be current on the research.

Ulam, developer of the Monte Carlo Method of solving probabilistic problems, was a mathematician, not a physicist, so Teller had run with the compression idea from there, Garwin said. But Teller still couldn't get it to work. Why did he then turn to Garwin, a twenty-three-year-old graduate student?

Physicists joke that their peak is in their twenties. After thirty, their creativity diminishes. Einstein was twenty-six when he published the special theory of relativity. Also, it was an easy assumption that if you were a protégé of the great Enrico Fermi, you had a brain. Thanks to praise from Fermi, Garwin was already known to be special.

"Dick Garwin, even at his young age, was a fine physicist, later even a great one, combining, as did his mentor Fermi, deep understanding of the physics of a problem with the facile and ingenious embodiment of an appropriate experiment," wrote Los Alamos physicist Harris Mayer.[2]

Teller's frustration was growing daily. He had formed a working group called the Family Committee ("family" for a new family of weapons), but they soon abandoned the task. During the attempt to design a thermonuclear device he divided the group up into eight sub-groups, each assigned to coming up with a design. All eight appeared to fail. What Teller didn't know—and scientists at Los Alamos found out later—was that all eight groups had made a series of errors that, had they been corrected, would have produced a hydrogen bomb. *All eight would have worked.* Teller got closer than he had ever imagined.

One delay in getting a final design for the bomb during the spring of 1951 was work on the Operation Greenhouse series of tests in the Pacific, scheduled for June, which was taking up resources.

"When I returned to Los Alamos in May 1951," Garwin wrote, "there had been some changes since my departure from there August 1950. Teller was full of enthusiasm, limited only by difficulties in recruiting the 'top team' to work on the new idea."[3] The potential plans for the bomb were down to two, one from Teller, based on work done with Ulam, and one from Fermi, according to Michael Bernardin, deputy associate director at the Associate Directorate for Weapons Physics at Los Alamos, who has seen the classified documents.[4] Both had issues.

There were three candidates for the material that would produce the fuel—deuterated ammonia, lithium deuteride, and liquid deuterium. The scientists picked one of the others but then return to the liquid deuterium. It is relatively cheap, contains vast amounts of energy, and is stable. The second problem was the geometrical configuration of the bomb: how big it was, what it looked like, how thick the enclosures were, what temperatures were needed and where. Third was the problem of the components, exactly what went where in the device.

When Teller met with Garwin, he told Garwin to look at papers in the library and to talk to the people who had worked on the bomb, including the engineers who would have to build it and its components. He asked Garwin to produce "a working diagram of a thermonuclear device." A "working diagram" is not a weapon, it is an experiment, Bernardin said.[5] The purpose was for this idea to be passed around the room as a topic of discussion, to present something to talk about.

"Teller wanted me as an experimenter to devise an experiment that would be absolutely persuasive that this would really work because the classical Super had just gone along and nobody could show it would work—or, well, wouldn't work—and it would be a big thing to try. You could always make it work by putting in enough tritium, but we didn't have the tritium. . . . I decided that I couldn't devise a little experiment these things don't scale very well," Garwin said.[6]

"I was an experimenter, but I knew enough about the theory that could be useful there," he said.[7] "So I just designed the whole thing, making all of the choices as to parameters, thickness of various walls, bolts, and the like. Because my work in experimental physics dealt not only with nuclear reactions but more helpfully with cryogenics in the form of liquid hydrogen and liquid deuterium and the vessels to hold them under various conditions, I just designed the whole thing. . . .[8]

"I said to myself, 'The best experiment is really a demonstration. If you try to make this [model] too small, it's not going to be persuasive. If you make it too big, it's going to cost a lot of money. Let's see what it would take to make a real thermonuclear weapon.' Because that's what you want to know. After you know how to make the weapon then you would devise an experiment that would prove the principles of the weapon. I took everybody's ideas and went away. . . . I just worked on them for a week or two and made the design choices. I talked to people and they gave me their strong views. . . .[9]

"I have a primary," he said, referring to a fission device. "I have a secondary [fusion]. I have a radiation case and what materials should these be and what thickness should these be. How do you put it together? The simplest fuel was liquid deuterium in principle. Since I was a low temperature physicist for my particle physics experiments I built liquid hydrogen and liquid deuterium dewars (thermos bottles) with windows on them so the beam could go through."[10] In a crude way, a hydrogen bomb is an extremely complicated explosive refrigerator because the fuel has to be brought to a very low temperature.

"It was duck soup to design this whole thing so that it would satisfy the nuclear properties and keep the primary bomb warm while the deuterium was cold and stuff like that."[11]

He even predicted the yield, how big the explosion would be, Bernardin said.[12]

"There is a whole chain of things that happens," Garwin said. "A nuclear explosion goes off so you have 10 kilotons of energy released in a small metal sphere of plutonium or uranium 235, and it's surrounded by a high explosive. What you need to do is put this thing in place so that the energy that comes off after a little while, I mean hundreds of a microsecond or so ... leaks out of the metal in the form of soft X-rays. ... If it came out in the form of ordinary light I'd have an aluminum mirror and focus it on something and the secondary will get hot. Here it is the energy and the pressure that are equivalent."[13]

But the complexity seemed overwhelming, even to Teller.

"Will the radiation leak into the walls too rapidly? Will the walls blow apart in this original box? Will the pusher break up as you push it in? Will the deuterium compress uniformly?" Garwin pointed out. "I said I'll work it up."[14]

Garwin published the design in a six-page "memo," five pages and a diagram, and, along with Fermi, presented it to the Theoretical Megaton Group, the main committee engaged in building the bomb, which had merged more or less with Teller's Family Committee. Hans Bethe chaired. Teller had not bothered to show up. The title of Garwin's paper is "Some Preliminary Indications of the Shape and Construction of a Sausage, Based on Ideas Prevailing in July 1951." The "sausage" was the body of the bomb.

Twenty-five people jammed in a small room when he presented his paper.

The meeting was called to order at 1:30 p.m., but, as Harris Mayer noted, "order didn't mean much to the scientists who had come together at that time." They continued to chat among themselves, sometimes about work, sometimes about something cute that their children had done. Bethe droned on. Some of the scientists continued working while he talked, fixing a recalcitrant equation or two, paying little attention.[15]

There was some discussion. The meetings were notorious for being indecisive and disorganized. There were frequent failures to communicate, but when their time came, Garwin and Fermi presented Garwin's design and were explicit in defending it. Bethe, already one of the giants of modern physics and an eventual Nobel Laureate, took issue with Garwin's idea for the thickness of the outer casing.

"You're wrong, Hans," said the twenty-three-year-old postdoc, "and here's why." When he was done, Bethe agreed he was wrong.[16]

The content of Garwin's paper is still, after all these years, secret. It is the crown jewel of the nuclear age.

Garwin's design would replace what was called the Classical Super. It was met with muted enthusiasm, but the group accepted the design and went on to other business. After the meeting, Garwin, Fermi, and Mayer walked to another building. Mayer and Garwin talked about the growing ideas of the bomb; Fermi was uncharacteristically quiet. Mayer and Garwin were upset because the meeting had seemed unenthusiastic about the paper and no one seemed to be in charge. Finally, Fermi stood up and said, "What we now need is a king." Someone had to make a decision.[17]

Garwin's design gradually became the project. It now was clear they could build a hydrogen bomb, and the mood at the laboratory changed dramatically. Teller may have believed, after all the struggles and conflict, he had been vindicated.

The best way to envision Garwin's design as an object is as a steel container, twenty feet in length and seven feet in diameter—a fat sausage. At one end was a fission bomb that would start the process. Just under the walls of the steel container was a layer of polyethylene that could channel the radiation. Inside that there was about five tons of natural uranium. Further inside was a dewar containing liquid deuterium, the fuel itself, chilled down to just twenty-four degrees above absolute zero, the temperature at which all molecular motion virtually stops. In the center was a rod of plutonium 239. Some reports indicate that within the plutonium was a pencil-thin space containing a deuterium-tritium mixture, a booster known as the "spark plug," essentially a second fission bomb. Gamma radiation from the primary heats and compresses the deuterium and tritium to fuse and burst with energy.

The primary fission bomb explodes. Running ahead of the rapidly expanding material, the radiation from the explosion vaporizes the polyethylene, creating a very hot plasma, adding to the pressure by expanding. The pressure pushes out toward the steel container and inward on the fuel. It quickly compresses and heats the deuterium, causing the plutonium to go critical and start a chain reaction. The plutonium ignites the deuterium, which produces a flood of neutrons that causes fission.

All of this happens within microseconds.

Then the whole thing blows up. Throw in all the plumbing required to keep things moving and to keep the deuterium liquid cool and you have a huge, heavy, fantastically complex device—Garwin's design.[18]

Garwin's more simple description: "I don't think you can say anything except it was a cylindrical design and a very, very big high explosive [that] used an awful lot of highly enriched uranium. But it's still classified, and [it] has never been revealed exactly how the deuterium-tritium was heated and compressed by the fission explosion. So I can't talk about that."[19]

In 1979, the *Progressive* published an article describing the hydrogen bomb along with diagrams that came close enough to the real thing that the government tried to get the federal courts to exercise prior restraint to block publication. This was after the Pentagon Papers incident when the US Supreme Court had already ruled it very difficult for the government to get prior restraint to prevent publication. The article contained numerous inaccuracies, but some critics at Los Alamos said the court action to deny the government's petition gave away "more secrets"[20] than the original article.

Even after it was accepted and a test scheduled, Teller wanted to tinker with the design. "It was one of Teller's problems. Nothing was ever finished," Garwin said.[21]

One small problem was that his design was "undeliverable," meaning it would not fit in a bomber; it was a vertical device, and bombers are horizontal. But with a modest redesign he soon had a version that was horizontal and would fit in a B-36, and several such bombs were built. That project was called Jughead after a character in a popular newspaper cartoon strip.[22]

Two Greenhouse tests had been scheduled for May 1951 at an atoll in the Marshall Islands, Eniwetok, so scientists could learn more about the possibility of a thermonuclear burn. The tests were scheduled before Teller and Ulam came up with their breakthrough; mostly the scientists were trying to see if they could make a bigger blast with less of the rare fuel than those used against Japan. The first test was called Greenhouse George, *G* for the seventh test.

A large fission bomb (perhaps the biggest bomb built to that date) was designed to ignite a small quantity of liquefied deuterium and tritium. George was a tower shot, a cylindrical device atop a steel tower. It was fired off on May 9, 1951, and produced a fireball almost 2,000 feet high, vaporizing the tower and several structures nearby, including 283 tons of equipment on the tower. It blasted a crater deep into the coral. George produced the first thermonuclear reaction ever achieved on Earth, far more powerful than the Hiroshima bomb. The next test, Greenhouse Item, was a spherical device with a Nagasaki-like bomb and the fuel—deuterium and tritium—in the center, but this time in gas at room temperature, and it didn't need complex cooling. That process was called "boosting," one of Teller's ideas. It had twice the yield of George. The test was a prototype of the Classical Super.[23] It incidentally validated the notion of compression.

Garwin worked on both tests, and one of his detectors was used in George. "I was much involved in 1950 talking to the people who were doing diagnostics on Greenhouse George and understanding how that was supposed to work," he said. "I talked to people from the Radiation Laboratory at Berkeley and Naval Research Laboratory because they had to build the apparatus. . . . I devised a couple of techniques myself." He even published a paper on the radiation.[24]

Despite the success, there still was a Teller problem. He felt that he had been shut out by other participants at a meeting in June at the Institute for Advanced Study in Princeton. His impatience grew. Greenhouse proved compression would work. Teller was now spending considerable time in Berkeley building a competing laboratory.

The first hydrogen bomb was to be called Ivy Mike. The design was frozen by the committee on January 18, 1952. Teller wanted the first test in July of that year, but in March problems rose about the design, mostly involving the channel the radiation was to flow through. Instead, the committee pushed back the date to November. Had they followed Teller's wishes, the test probably would have failed.

Little of the device was assembled in the US. Parts came from everywhere in a supply chain to the Pacific. The steel casing came from a company in Buffalo, New York, the dewars from Boulder, Colorado. Everything had to be shipped to Eniwetok, 3,000 miles west of Hawaii. The indigenous population was removed to other islands. Several other tests, including Greenhouse, were staged on the atoll, and some of the buildings for those tests were restored for Mike. People started moving there by March, and by October, 9,000 military personnel and 2,000 civilians, including some from companies that had helped construct portions of the device, had moved in, along with hundreds of ships and aircraft, including an aircraft carrier, barges, motorboats, and helicopters. The men (and they were all men) lived on shipboard and in tents on the island itself. The US Navy moved in mountains of supplies to feed everyone, and everyone was very well fed.[25]

Mike (known to the scientists simply as the "sausage") was to be tested on Elugelab, an island north of Eniwetok. Thirty adjoining islands were also drafted to house scientists and technicians. A large mound was built on Elugelab so the line of sight would be improved, and a six-story, open air structure was placed atop it to house Mike. A 9,000-foot-long plywood tunnel was built so that radiation from the device could be measured over the first microseconds after ignition to see how the ultra-rapid cascade of events went. It was raised so that it provided a line-of-sight without distortion by the curvature of the Earth.[26]

Assembly of the bomb began in September. Moving the radioactive materials, including a few grains of priceless tritium, was particularly tricky since some of it can ignite spontaneously when exposed to air. Mike was loaded through a manhole in the casing—very slowly. They were finished the night of October 31, and the arming team left by ship to a point ten miles away.

No one knew how big the explosion was going to be—if there was going to be an explosion. At 7:15 a.m., November 1 (October 31 back in the States), a radio signal was sent from a ship and ninety-two detonators on Mike fired simultaneously. Within microseconds it produced a fireball hotter than the interior of the sun that grew to three miles across. People thirty miles away could feel the heat. Orange, purple, and red, the fireball rose more than 100,000 feet, hit the stratopause, and spread out in mushroom formation, a huge, virulent canopy that stretched more than one hundred miles. Elugelab was vaporized.

The yield was 10.4 megatons, a thousand times more powerful than the bomb that destroyed Hiroshima.[27] The world was a different place.

Teller, still bitter about how he was treated, had quit Los Alamos after he was rejected as leader of the hydrogen bomb work and was off in Berkeley. He said he was busy that day. He watched a seismograph, and when Mike exploded he saw the seismograph jump and knew there was a bomb. The blast was that powerful. He was perhaps just a little peeved that Los Alamos, which had resisted his idea for years, had succeeded.

And Richard Garwin?

He still had a month to go at Los Alamos after he designed the bomb and spent it designing the flyable version. He knew what he'd been asked to do was create an experiment, so it didn't matter that Mike stood vertically. A bomb had to fit in a bomber. Then he left New Mexico.

"I was in Chicago," he said, explaining where he was when Mike went off. He had no interest in seeing if his design worked. In fact, despite a life of working to remove nuclear weapons from the Earth, he has never seen one go off. "I have a good imagination."[28]

Unknown to him, five or six of his flyable Jughead bombs were actually built, and for a while they consisted of the entire thermonuclear arsenal of the United States. Bernardin says they were never deployed.[29]

Teller spent most of the rest of his life taking credit and doing everything he could to make sure he got credit. He participated, he told everyone. Ulam did not. He never mentioned who actually solved the problem and designed the weapon until 1979, when he had a heart attack. Teller called in a friend, George Keyworth II, and recorded a "last testament" to straighten out the controversy and finally kill any idea Stanislav Ulam had a major role in the bomb. Keyworth would eventually leak it to the *New York Times*.

> In the early 1950's when I had the first crude design of the hydrogen bomb, Dick Garwin came to Los Alamos and asked me how he could help. Actually the design I had in mind was not that of a real bomb but of a model for an experiment. I asked Garwin to change this crude design into something approximating a blueprint. He did so in a short time—a week or two. That experiment was carried out. Garwin's blueprint had been criticized by many people, including Hans Bethe. In the end the shot was fired almost precisely according to Garwin's design, and it worked as expected.[30]

Not quite.

It was, Bernardin said, the best representation of a hydrogen bomb. It was the starting point for Mike, not Mike.[31]

Bernardin, who has seen Garwin's paper, said if you pasted the Garwin diagram on a wall and then pasted a similar one from Mike, you could see the significant differences between the two. The final diagram showed all the work done by others and several important issues that had been resolved by others. On the other hand, the altered version was based on Garwin's final diagram. It was from that altered paper Mike eventually was built.[32]

Kenneth Ford, a Los Alamos physicist whose book, *Building the H Bomb: A Personal History*, the government wanted to censor, wrote that Garwin "laid out a design with full specifics of size, shape, and composition for what would be the Mike shot fired the next year."[33]

No one except the people involved knew of Garwin's association with the hydrogen bomb. Richard Rhodes, who wrote the definitive history of the bomb, missed it because no one told him about it, including Garwin. Almost none of the reference literature mentions it. Garwin never even told his father. His family did not find out until years later when they read about it in a story written by William Broad at the *New York Times*.

"And they built it just as I had designed it. And it worked," Garwin said.[34]

Garwin said that at the time he was not aware of the GAC report that recommended the bomb not be built because it was classified, even less the Fermi-Rabbi report. Does he have a moral issue with what he did?

"I didn't think about it. Maybe had little bit of the attitude of Edward Teller. You work these things out first, you show they can be done, and then you decide not to use them. I was just working on the problems and seeing whether I could solve them."[35]

"Do you feel like you are actually emotionally somewhat unaffected by it?" he was asked.[36]

"If I don't do these things somebody else will do it. It may take a lot longer; it will certainly cost a lot more."[37]

He would, however, spend the rest of his life trying to make sure no one ever used his device.

CHAPTER FIVE

GARWIN, LEDERMAN, AND THE MARX BROTHERS

Back in Chicago, Garwin was unhappy. His family was eligible for university housing and moved to East Fifty-Sixth and South Ellis just off campus, a half block north of his office in the Institute for Nuclear Studies, now known as the Enrico Fermi Institute. He gathered friends, many of them, like Harold Agnew and his wife Beverly, and Murph Goldberger and his wife Mildred. They became lifelong friends, and many of the guests would become the icons of mid-twentieth-century physics. There were dinner parties every Saturday and birthday parties for the children.

Lois apparently recognized that doing science, particularly at that part of Richard's career, was a social phenomenon. They talked about who they should invite for dinner. Space was limited, so if they entertained, they would have a limit of six. Dinner was buffet style, eating from trays.

"All those people," Lois remembered. "I even had the nerve to invite them to our house for dinner—when I say 'our house,' our apartment, where the only table I had was a Formica and chrome table in the kitchen. The kitchen was a large kitchen, and at one end of it, there was room for a table and chairs. I remember very well having the Fermis—I think it was the Fermis and Tellers—and maybe one other couple—to dinner ... and I served them spaghetti with meat sauce," she said with a laugh. "How I had such nerve, such *chutzpah*. Terrible. I blush to think of it."[1]

They were close to the Tellers—Edward and Mici (pronounced Mitzi). Lois described him as "very affable" and "gracious." Lois was especially fond—as was everyone else—of Fermi and Laura. "She was an absolutely marvelous woman ... generous and loving." She described Enrico as humble, without airs. "He wasn't the grand professor." They exchanged dinner invitations. Seventy years later, memory of the affection lingers.

The problem was to be expected: the men sat around and talked physics, leaving the women to their own devices. So the women devised strategies, such as making sure that if two men were talking, a woman would sit down between them. Two men on a sofa? A woman would head for the middle. Or pull up a chair. "You shouldn't let them get off on their own," Lois said.

One physicist was taciturn in social events. The women learned he would loosen up with a drink, and, Lois thought, he'd become quite charming. But since he didn't like to drink they had to serve him in a way that was not obvious.

Garwin's work filled his life, but he always took time to explain what he was doing to Lois, and she was able to follow.

"I began to feel sorry for my women friends whose husbands did not do that—you really do get shut out, no matter how hard—no matter how much else you have in common, whether you play games, or you appreciate art or music, or whatever, or you travel. But if the core of your life isn't shared, then it gets harder and harder I think, rather than easier," she told Dan Ford.

One tradition was a faculty-student songfest, in which everyone was serenaded with melodies, all around the same theme. One year, just before Garwin left, the theme was "No, never mind, I'll stick to my Nobel Prize." The song for Garwin went:

> There's a special deal, said General Joseph Meyer,
> For those who die under Korean skies.
> Oh, no, thank you, said Colonel Richard Garwin,
> I'll be content with a Noble Prize.[2]

They remained close to their families in Cleveland. The Garwins back in Ohio eventually prospered. "They considered themselves lucky," Lois said about the Cleveland Garwins. "They started with nothing. . . . Both their sons have PhDs. Both of them have been very successful. They themselves were able to travel. They visited us in Europe when we lived in Switzerland. They did sort of a grand tour. They took cruises to the Bahamas or the Caribbean. . . . They did a cruise in the Pacific or something, to Hong Kong, and . . . they spent a lot of time in Japan on that trip. They really considered themselves extremely fortunate. They were among the blessed."[3]

Still, not everything was copacetic in Chicago. Things were not going particularly well in the lab. Garwin and one of the other scientists in the lab were having a personality clash, Dan Ford reported, and the best solution would have been for Garwin to leave. There were other issues.

"Chicago was rife with crime, to the extent that I would visibly carry a large pipe wrench with me as I went back to work at the Chicago synchrocyclotron after dinner," Garwin said. Sometimes he did not come home until two in the morning. "That was a primary reason for my decision in 1952 to seek employment elsewhere, despite the attractiveness of the physics faculty at Chicago," he said.[4]

"I guess I visited Berkeley, but IBM had a laboratory since before the war really. It was in Columbia University buildings, founded by Wallace J. Eckert, an astronomer who introduced the punched card to scientific computing. And now in 1952, having gone into scientific computing in a big way at the Watson Scientific Computing Laboratory in 1945 on 116th Street, they wanted to move from vacuum tube computers to solid state physics. Eckert, a very farseeing person, was a faculty member in the Columbia University Department of Astronomy and good friend of [Isidor] Rabi. They decided that solid state physics was a good thing to do. There wasn't any at Columbia University, so the Watson Laboratory was going to have a strong push into this.[5]

"They had, through contacts, asked Emilio Segre to come see whether he would like to head this laboratory. He was at the University of Illinois on sabbatical from Berkeley that year, which was a time of problems in California with the loyalty oath and so on. I went down to Illinois from Chicago to see Segre. It turned out that Segre thought this was a very good idea, so I visited Wallace Eckert in New York. Eventually Segre decided not to [take the job] and I decided to join the laboratory, so Wallace was director for a while and then I was director."

The arrangement was perfect. Garwin had already established himself as a consultant for the government, so the offer was that he would have one-third of his IBM time free to work for the government. It was in his contract.

"Garwin went to work for IBM and I think in a short time he was irreplaceable," Leon Lederman said. "I remember once the CEO of IBM was looking for Garwin. He called him up and his secretary said he's out of town. 'Well where is he?' 'I don't know. He didn't let me know.' You think the CEO would be a little irritated. Instead a very polite note was sent around to all the scientists saying, please when you're leaving town let your secretaries know where you are just in case you're needed. So Garwin organized a protest against this restriction on his freedom. It was pretty amazing."[6]

Explaining even in general terms what Garwin proved in his most famous experiment is difficult. The details are mired in quantum mechanics, a field of science in which language often confuses reality. Quantum mechanics is the dream stuff is made of, said one physicist.

One of the glorious attributes of physics is the way physicists use language, especially English. To a physicist, a magnetic moment has nothing to do with suddenly falling in love across a crowded room. And what other science would devote itself to spending millions of dollars searching for a naked bottom quark? But use of language, like most of logic, falls apart, when talking and writing about quantum physics. It's the reality of Schrödinger's Cat and where Einstein scoffed that God doesn't play dice with the universe. A modern physicist might argue that He does and the Garwin-Lederman-Weinrich experiment is evidence.

With that in mind: suppose you stood in front of a mirror and waved your right arm. The image in the mirror would raise his or her left arm and wave back. Always, and it was forever thus. In physics, that's called *parity*, a kind of symmetry that everything you did would be done in the mirror world exactly the same way only in the opposite direction, and it would look perfectly natural.

If you dropped a ball in front of the mirror, your mirror image would drop a ball and both would land at exactly the same time. That's because the laws of physics in the real world also apply in the mirror world. It is always true. That's called *conservation of parity*, parity always happens.

Let's say you were in a communication with an alien planet. You can assume that right-handedness means the same to them as it does to us, although if they were a mirror universe, left would be right and right left. Yet the laws of physics in both worlds would be exactly the same. Always.

You photograph someone doing something, and then you photograph the mirror image of that person doing the same deed. If you showed the picture to another person it would be impossible for that person to tell if he or she was looking at a photograph of the person acting or the mirror image of that person acting because the two would be exactly alike in all ways. It would always be the same, only the opposite.[7]

"It turns out," Garwin said, "it doesn't work that way. The mirror world does not evolve the same way as the world."[8]

The best way to explain what happened would be to call in the Marx brothers. Think of the mirror scene in the movie *Duck Soup*. Groucho sees an image of himself in a mirror that reacts exactly as a mirror should, duplicating every movement. It's really Harpo disguised as Groucho. Harpo broke the mirror and doesn't want Groucho to know, so he is imitating the mirror by reproducing what Groucho is doing exactly, even the silly walks. Parity. Both images move exactly the same way. Suddenly, Groucho spins, but Harpo, the image in the mirror, just stands smiling. *Violation of parity*.

Parity seems to work in the "real" world we know and in the

quantum world that keeps physicists up at night. It is especially true in the world of subatomic particles.

What Garwin, Lederman, and Weinrich proved was that in certain circumstances subatomic particles will violate parity. The mirror world can go its own way. This proof destroyed a theory held dear by physicists for thirty years, opened up a whole new understanding of chemistry, and incidentally opened a new window for science fiction.

Keep in mind this is quantum mechanics, and as Richard Feynman once said, "If you think you understand quantum mechanics, you don't understand quantum mechanics." As an example, Feynman described a scene in which a driver was stopped by a police officer.[9]

"What's wrong, Officer?" the driver said.

"You were speeding."

"No, I wasn't."

"Yes, I clocked you at sixty miles an hour," the police officer said.

"Aha," the driver said. "That's crazy. I've only been driving for twenty minutes."

And that is quantum theory, Feynman said. If you don't understand it, you got it right.

In 1848, Louis Pasteur found that you can have two forms of the same chemical, what are called isomers. The only difference was that they rotated light in opposite directions. Polarized light sends the vibrations of the photons of light in one direction or another—horizontal or vertical or rotating right or rotating left, what physicists now call handedness. You have right-handed molecules; you have left-handed molecules. One is the mirror image of the other, like a right-handed screw or a left-handed screw. Living things would use one or the other, never both. Inorganic nature produced equal amounts of each. That made sense to physicists who have an innate affection for symmetry. There is a balance, a *feng shui* to nature.

In the 1950s, observations demonstrated that the strong interactions among particles led to the conclusion that parity was also conserved in the strong force, the force that keeps atomic nuclei together.

As for the weak force of nature, it was simply assumed, with very little data, that this conclusion would apply there as well. (Gravity appeared to have nothing to do with this.) Physicists couldn't imagine a nature in which this conservation of parity did not apply everywhere. That belief held for thirty years. All of this came to be called part of the Standard Model, the explanation for *everything*.

Enter two Chinese-born scientists, Tsung Dao Lee and Chen Ning Yang. Both men had been studying at the University of Chicago— Lee under Fermi, Yang under Teller. Garwin knew them at Chicago and described the two as theorists, not experimentalists, a distinction important in the physics community, where each group tends to look down on the other.[10] They both ended up at the Institute for Advanced Study at Princeton, down the hall from Einstein. In 1956, they decided to clear up a conundrum, the so-called Theta-Tau Puzzle.

Cosmic rays engender a whole menagerie of particles, one of which disintegrated into three other particles, called pi mesons, or pions. It was named the tau meson. Yet there was a similar particle that disintegrated into just two, the theta meson. They had the same mass and lived the same length of time, measured in nanoseconds, and were, in every other way, identical, except that the decay products had different parities, or, alternatively, the particle did not have a specific parity. They had to be different species of particles. If they were in fact different faces of the same particle, parity had to have been violated.

In April 1956, at a conference in Rochester, New York, Lee and Yang suggested that was so, that certain elementary particles might occur in two forms with different parities. Richard Feynman and his roommate at the conference, Martin Block, said that maybe they were right, that under certain circumstances, parity is not conserved. Feynman admitted he didn't really believe that and even made a $50 bet it would turn out wrong. "I thought the idea (of parity violation) unlikely, but possible, and a very exciting possibility," Feynman said. Wigner, also at the meeting, however thought maybe they were right.[11]

Chatting over tea at the White Rose Cafe near Columbia University, Lee and Yang, still puzzled, realized that every experiment done so far involving the weak force avoided testing for parity conservation. Experiments on the strong force and electromagnetism seemed unshakeable, but there was no evidence from experiments in the weak force that parity was conserved. New thinking was needed.

"The fact that parity conservation in the weak interactions was believed for so long without experimental support was very startling. But what was more startling was the prospect that a space-time symmetry law which the physicists have learned so well may be violated. This prospect did not appeal to us," Yang said later.[12] They published a paper in *Physical Review* suggesting that experiments were needed to determine whether weak interactions can tell the difference between right and left. They proposed several.

Enter a third Chinese-born experimental physicist, Chien-Shiung Wu, a woman professor at Columbia University and an internationally known expert on beta decay, an aspect of the weak force. One of the experiments Lee and Yang suggested involved the spin of the cobalt-60 atom. If someone polarized the nuclei of those atoms with magnetic fields—a magnetic moment—it would align the spin of the nuclei in one direction or another, from the north or the south poles. Beta rays would be emitted from both poles of the atoms, but the north and south poles would be reversed in a mirror image. If the rays were equally distributed north and south, parity would have been conserved. If more came from one pole than the other, the theory would be wrong because the poles should be reversed under parity inversion—just like a mirror.

Wu wanted to test the theory or run the experiment. Needing expertise and equipment for the cold temperatures needed to do the experiment, she turned to the National Bureau of Standards, a government facility with the equipment and expertise needed. On December 27, 1956, Wu and her colleagues succeeded, showing parity had been violated.[13]

Lee reported the finding to his colleagues, including Leon Lederman, who worked with Columbia's cyclotron. An issue remained. Was parity really violated, or was there something wrong with the experiment? Wu was having difficulty replicating her success.

Wu's experiment was physically difficult, requiring temperatures of one-hundredth of a degree kelvin, one-hundredth of a degree above the absolute zero, and it had to be held in the device for a very long time. Wu continued to work but kept running into difficulties, laboring nights and weekends, sometimes sleeping next to the device. Eventually the researchers worked it out, and on January 9, 1957, they toasted the death of the law of parity with a bottle of Chateau Lafitte-Rothschild, 1949.[14]

Parity was not necessarily conserved in the influence of the weak force.

Leon Lederman thought he could take another of Lee and Yang's suggested experiments and independently prove that parity is not conserved.[15]

Muons were first detected in cosmic rays by scientists using Geiger counters. If the scientists went to a place that was shielded from cosmic rays, deep underground or underwater, the clicking would slow down in the counters but never completely stop. After the normal radiation count dropped off, something was still getting through. Those turned out to be what became called muons. They weren't the primary cosmic rays; they were caused when the primary rays, the protons, produced nuclear interactions in the atmosphere and created unstable particles of mass somewhere between that of a proton and that of an electron. First named mesotrons but now known as mesons, they could be imaged in special photographic emulsions.

"You could see a charged particle, a pion, coming in, increasing its ionization density because it was coming to rest. So then it stops and another charged particle comes out. This other charged particle has considerable energy, always has a fixed range, so many microns, comes to rest and it gives rise to another charged particle. That was the discovery of the muon," Garwin said.[16]

"Columbia University Physicist [Isidor] Rabi said about the muon, 'Who ordered that?' ... They didn't seem to do anything," Garwin said.[17]

Muons can go through almost anything without using nuclear interactions, but they lose energy to electrons and slow down and stop. They are deflected by heavy atoms such as lead or uranium. Muons, from the Columbia University cyclotron, traveling at a good fraction of the speed of light, will stop within a stunning eight inches of pure carbon or graphite, or an equivalent mass-thickness of wood or water.

Mesons had first been discovered by scientists exposing photographic silver emulsion film. The particles appeared in microscopes as tiny black lines on the film no one could explain. In the 1940s, physicists were producing them in cyclotrons at Berkeley, Columbia, Chicago, and elsewhere. The cyclotrons produced pi mesons, and the pi mesons decayed into muons either in flight or after coming to rest. Positive muons have a half-life of two microseconds after they stop in matter. No one knew about negative muons.

"We are all bathed in muons. Every square centimeter at sea level has a muon passing through once a minute, all the time," Garwin said. "Cosmic ray muons contribute significantly to the radiation dose received by humans at sea level—about twice as much annually as the dose received from the potassium-40 radioactivity built into our bodies with the stardust from which we are all made."[18]

The only way to avoid cosmic-ray muons is to hide beneath the Antarctic ice cap or hundreds of meters or so underwater or in deep tunnels. Muons are wonderful probes. They have even been used to "X-ray" the Great Pyramid of Cheops in Egypt.[19]

Garwin was in Poughkeepsie, New York, in an IBM lab the evening of January 4. He had invented a superconducting computer and was heading a team of 100 at three IBM locations. He had changed his specialty. He moved from particle physics to low-temperature physics in 1953 because, he said, "I didn't like the sociology of working in large groups and telling people six weeks ahead of time what we are doing." He had been specializing in helium frozen solid.[20]

Lois answered the phone when Lederman called Garwin's home. Lederman asked her to have Garwin call him back as soon as he arrived. Garwin did, and Lederman gave him the news that Wu was finally getting consistent results. He said he had an idea to use the cyclotron to perform another of Lee and Yang's potentially useful experiments, involving the polarization of the muons.

"Leon and I agreed to meet at the Nevis Cyclotron of Columbia University in fifteen minutes, at 8 p.m. Necessity—and even adversity—being the mother of invention, we benefited greatly from the fact that the machine shop and the stockroom were closed and that the cyclotron would begin its weekend rest Saturday morning. Leon and Marcel Weinrich, his graduate student, had a setup in the external beam of the cyclotron, investigating the decay of muons, plus and minus," Garwin said.[21]

"The cyclotron was going to shut off Saturday morning for its weekend maintenance and turn on Monday morning, typically," Garwin said. "But there was more maintenance to do this time than usual, so it wouldn't start going until Monday evening."[22]

They had beams of muons coming out of the cyclotron from the decay of pions in flight within the cyclotron. The cyclotron would produce the pions, shielded by five meters of concrete. When the concrete was cast, a small hole was created a couple of inches in diameter, straight through the wall. The mixed muon and pion beam would come out of the cyclotron and go in a straight line into the graphite block. The particles lose their energy and stop. Four inches of carbon block would stop the pions; muons would stop in six inches. A thin counter gave a light pulse when a muon came in. There would be another counter that made sure nothing came out. Measuring one counter to another, you could tell when a charged particle stopped in the graphite.

It stops in a very short time, less than a nanosecond.

The lifetime of the positive muons were well known. The negative muons were what Lederman and Weinrich were working on.

The goal of their experiments was to measure the exponential decay of the muons. Weinrich was having trouble because the negative muons in his experiment were not decaying exponentially.

After Garwin arrived, he suggested applying a magnetic field to the graphite block, continuously rotating the muons' magnetic moment so the scientists could measure the polarization.

"The point there is that this advance took two ideas," Garwin said, "and I didn't have them both. But the two of us working together had them both. One was that the muons were already polarized with their spins along the direction of flight, and the other that you rotate their spins rather than looking in detail at the direction of the emission of the electrons. That's a little complicated to understand. The more theory you know, the harder it is to understand."[23]

"The whole experiment took thirty-six hours," Lederman said in his version of the story. "Probably if not for Garwin, it would have taken maybe thirty-six days. But he had some ideas on how to make the experiment very quickly. . . . We talked about it. There was a graduate student [Weinrich] there puttering around with some of his thesis experiments. And we said that's just the experiment we needed. So we took apart the student's experiment, the student started crying. We said, don't cry, everything will be all right. Then in a day and a half we had a new experiment setup."[24]

Garwin found a Lucite tube six inches in diameter and eight inches long in the workshop's trash bin and went to work on a lathe. They got a power supply. A couple of hours later they plugged in an analyzer Garwin had brought from Chicago that had been unused by the Columbia lab and was sitting around on a shelf. Wires were rearranged. They took data for twenty minutes, changed the magnetic field, and collected information this way for several hours, fine-tuning the intensity of the magnetic field. They persuaded the staff there to keep the cyclotron running past its weekend shut-off time.

But in the final hour the count no longer changed as the intensity of the magnetic field was changed.

"That was disappointing. You are getting data and then it goes away. We didn't understand that," Garwin said.[25] When the cyclotron shut down they found that the wire coil on the Lucite had fallen off.

Monday, Garwin wound another coil in a sturdier way, directly on the graphite block. By 3:00 a.m. Tuesday morning, after Lederman went home, Garwin got the results they were looking for. He even wrote the draft for the scientific paper they were going to get published.

"I called him up and told him it was Stockholm calling."[26] Stockholm is where they award the Nobel Prize.

Lederman returned, and by noon the next day Lee and Yang came to the cyclotron to see what they had. Of the electrons emitted when the muon decayed, they found twice as many electrons going out the back end as going forward.

"That said that parity had been maximally violated," Garwin said.[27]

"The mirror world is a perfectly feasible real world," Lederman said. "In our experiment, we found a particle, whose mirror image doesn't exist. There are ideas in which the muscles of the brain haven't been used along those lines." It had been inconceivable.[28]

"The mirror image of the world doesn't behave the same way as the world itself," Garwin said. "That was a front page article in the *New York Times*. Most things are bilaterally symmetrical in people, most things, and the things that are unique are mostly on the center line. There are a few things like the heart. There's only one of it and it's not on the center. But it would work exactly the same way. If you made a physical copy of the world and it were reversed, the way you see the world in the mirror, it would evolve in just the same way the world does."[29]

Right-handed screws in the mirror image would be left-handed. All the right-handed nuts would be left-handed. And everything would work exactly the same way, the theory went.

"It turns out it doesn't work that way," Garwin said. "The mirror world does not evolve the way as the world itself. Reflection invariance is broken. . . . It really meant that world wasn't at all the way we had thought."[30]

News of what Garwin, Lederman, and Weinrich did got out quickly, but they felt obliged to wait for Wu to publish, and she wasn't ready yet.

"Garwin and I waited a very painful week until Wu had made her last tests," Lederman said. The two papers were published simultaneously in *Physical Review*.[31]

The publication stirred some controversy. A third team, led by Valentine Telegdi, had completed a related experiment with similar results, but Telegdi's paper was missing from the review. There was a history of some animosity between Telegdi and Garwin. Telegdi found Garwin sometimes difficult.

"Telegdi said that Garwin was almost impossibly abrasive and said he would be standing in the laboratory and he'd like to look over other people's shoulders at what they were doing. . . . And make comments about how they could do it better," Murray Gell-Mann said. Telegdi said Garwin would start the comment with "Only an idiot would . . ."[32]

Gell-Mann thought Columbia University put pressure on the journal to protect Lederman and Garwin. Someone in the Columbia team had called Chicago to see how Telegdi was doing and was accidentally misled to believe Telegdi wasn't close, Gell-Mann said:

> The one of Garwin and Lederman was much more clear cut and simple but they both gave the same result. But the politics of Columbia University was such that they managed to get the *Physical Review* which is under the thumb of Columbia University to publish one and not the other. The other one was put off until the next issue. Valentine was furious. . . .
>
> So they used their influence to get the Chicago paper put off until the next issue of the journal. It was very sad. Valentine was furious and complained and finally in response to his complaint they published a little paragraph saying that article was published in the following issue for purely technical reasons which made it sound as if it had something to do with the equipment. Of course, it didn't have anything to do with the equipment which is just photographic plates. It had to do with the fact that they pushed to get their article in and the other one not. So, anyway, that's what happens. I'm sure Dick didn't have anything to do with that. Dick was not that kind of underhanded person but Lederman is.[33]

"That was a wonderful experiment because it kept me busy for a couple of years," Garwin said. "It gave me and other scientists a new tool for measuring the effects of the magnetic moment on these particles."[34]

Yang and Lee won Nobel Prizes, but Wu did not, some think for no other reason than that she was a woman. Garwin thinks that citing any experimentalist would have opened a can of worms because Wu worked with a group at the National Bureau of Standards in Washington, and the Nobel Prize cannot be shared by more than three people. That also precluded adding Garwin, Lederman, and Weinrich to the list.

"Many people said that they should've shared the Nobel Prize," said Walter Munk, an oceanographer who knew Garwin as a member of JASON, "but Garwin wrote in the paper that they were doing this work following the discovery by these other people. He didn't have to say that."[35]

"You could argue it was the biggest discovery in at least the last half of the [twentieth] century," said Norman Ramsey, who won a Nobel in 1989. "Because people assumed [conservation of parity] was there and it gave various kinds of problems, but it also discouraged them from looking for things. And once that was discovered ... well it did two things. People would search for a lot of the things they thought were obvious and weren't. But it made possible a whole bunch of ... explanation of how the radioactivities behave and so on, which didn't make sense before."[36]

The Garwin-Lederman-Weinrich experiment is still discussed in physics textbooks.

Garwin immediately quit the computer project at IBM to devote his scientific career to the world of research he had helped open.[37]

CHAPTER SIX
IBM AND LAMP LIGHT

Garwin had found a home at International Business Machines, IBM, then the paragon of American business. IBM and American Telephone & Telegraph Company (AT&T) had monopolies or near monopolies on technologies, helped in part because they spent much of their gigantic profits on research and development. Both companies ran research laboratories that have not been matched by any corporations in the world since. They could even afford to let their scientists play with theoretical or pure research—things that had nothing directly to do with their core business—both because one could never tell what might fall out, and because it seemed the right thing to do with their profits. Five IBM employees have won Nobel Prizes; eight prizes have been awarded for work done at Bell Labs. The lab, in Holmdel, New Jersey, designed by Eero Saarinen, remains one of the most beautiful buildings in the world, now mostly abandoned, and it's likely no corporation now could afford to build it. IBM has a Saarinen building too.

Garwin said he enjoyed the work he was doing at Los Alamos, where he spent his summers, and "I was good at it."[1] He wanted to continue with that work, and with his increasing time working for the government. He had begun working on defense matters.

So, they worked out a deal: Garwin would spend one-third of his time working for the government and the rest of the time at IBM's lab at Columbia University, where he had a courtesy appointment in the physics department. He stayed on full salary, since he was not being paid for his government work, and returned any fees he did receive.

"It required a lot of tolerance on their part," he said of IBM.[2]

One of the reasons for the move was to have collaboration. He had few colleagues at Chicago, only one graduate student, Maurice Glicksman, "whom I shamefully neglected."[3] Garwin was in Los Alamos every summer for two or three months and collaborated there. IBM offered him a chance to hire and work with teams.

Garwin only had half a dozen graduate students at IBM, and he neglected them as well.

"I was always away in Washington or doing something."[4] The few graduate students he had did well. Miriam Sarachik became president of the American Physical Society.

A second son, Tom, was born in 1953, just after the family moved from Chicago to New York. Two years later, looking for a home and better schools, the Garwins moved from the Bronx to Westchester County and Scarsdale. They had tried Hastings, New York, first— there were a number of Columbia people there—but it was small and they didn't find a house they liked.

Their daughter Laura was born in 1957, but by then, Garwin's peripatetic life had begun. Two days after Laura and Lois came back from the hospital, he was on his way to Europe, but had installed a new dishwasher in the house as a present. He, Leon Lederman, and T. D. Lee went to Venice to report on the parity experiment.

Sometimes Garwin took the family on trips. He took them to Geneva in 1958 for a conference on surprise attack—two adults and three children and skis for everyone. "When people are young, it's a lot easier."[5] His children still all remember the trip fifty years later. Laura remembers going to a Pugwash meeting in Romania, and trips to Cape Cod, many to La Jolla, Los Alamos, and England.

In the 1970s, the family began going to meetings held by JASON, an under-the-radar scientific organization steeped in defense work, first in Woods Hole, Massachusetts, and later in La Jolla.

"I always had plenty of time and solitude to do my work," Garwin said. "I've always done my chores around the house—helped with the dishes or the cleaning or whatever. Sometimes if the children at night would wake up and their diapers would need changing, I would do that, because I found it easier to wake up than my wife does. But she took care of the children and we always had dinner together in those days. But there was plenty of time to do my work; it's amazing how much time there was."[6]

The marriage with IBM lasted the rest of his working life. It was a compatible partnership despite Garwin's discomfort with corporate science.

> There's a problem with organizations which are created for a purpose and then are involved in controversial things with a small part of their efforts. Mostly it is prevented because people on the board of directors or board of trustees say it's our business to see that our principal purpose is carried out; that is, the stockholders' benefit and the customers get good computers and so on. So in addition to the three principles that Tom Watson Sr., I guess, enunciated for the company—that is, service to our stockholders, service to our employees, benefits to our customers—I encouraged people to add a fourth, and that is, to be a good corporate citizen, that is, to have service to society in general.
>
> Of course you can't do that to the extent that the company isn't profitable, but if you find some way in which it is cost effective to help the rest of the country or world, then you ought to do it. The calculus that I use to justify IBM's support of these activities was simply to say that IBM is more than one percent of the Gross National Product, and so if I can save a billion dollars here or there, then one percent—10 million dollars—of that is money that doesn't have to be collected in taxes from IBM. It's not that IBM benefits by selling computing machines to the people that I consult with. In fact it might be quite the opposite. I may never know. But it is this other. And so whatever rationale IBM uses—and there must be times when people complain to them, but I hardly ever see those complaints—they continue to do this, and I think there ought to be more and not less such activity.[7]

IBM was enormously profitable and could afford Garwin. Its stock and that of AT&T's were probably held by millions of widows and retired school teachers who lived at least partially on the dividends and the knowledge that nothing bad would happen.

Most of the time Garwin could not tell his bosses at IBM what he was doing on government business. He had a very high security clearance, and most of his work was classified.

He was still consulting with nuclear weapons and looked at nuclear testing. In 1950, he did a paper on nuclear "fratricide," limiting the destructive power of one nuclear weapon by exploding another one close by and quickly.

Almost as soon as Garwin got there, IBM asked him to go to Lexington, Massachusetts, to work on Project Lamp Light for a year, research into tracking Soviet bombers that might be coming in over the oceans to bomb America. It began his introduction to political science, with "political" as an adjective.

IBM had done earlier work on a related project in the early 1950s. It rose from a summer study program, a process very popular with the Charles River crowd. MIT thought that every problem could be solved in a (subsidized) summer study. Lamp Light went on well into the autumn. Garwin was beginning to consult not just with Los Alamos but with commercial firms such as General Dynamics.

"The company [IBM] really encouraged me—the only time they ever did such a thing. ... But there were interesting people working in a technical task-force mode, so I learned a lot from that," he said.[8]

He went to Cambridge, Massachusetts, three days a week for more than a year, flying to Boston on the Eastern Airlines Shuttle, at IBM Monday and Friday, and weekends at home, an apartment in Riverdale in the Bronx.

"This is not what I had signed up for at IBM, but I had intended to involve myself in the technology of information storage and transmission, so I was, in fact, quite interested in the subject," he said.[9]

At the start of the Cold War, the US had one great defense against the Soviets, a string of radar stations across northern Canada and Alaska, the so-called DEW (Distant Early Warning) radar line. The DEW line was set up to spot the Soviet bombers as they crossed into the two nations' airspace. It meant the inevitable air battle would be farther from the US than it would otherwise be, and give the American military more time to react. The radar data fed into the Semi-Automatic Ground Environment system (SAGE). Orders would fly out of SAGE's headquarters to launch planes and Nike surface-to-air missiles. The system was aimed at bombers coming in over the land, mostly across the Arctic.

The aim of the six-month-long Project Lamp Light at MIT's Lincoln Laboratory was to extend SAGE out to sea. A hundred military and civilian scientists, including Garwin, were involved.

The project was under the leadership of Jerry Wiesner and Jerrold Zacharias, two of the most well-connected scientists in the country, and they asked Tom Watson, IBM's head, specifically for Garwin.[10] Wiesner would eventually become John F. Kennedy's science advisor and chair of the President's Science Advisory Committee (PSAC). Zacharias also served on PSAC. The time at Lamp Light was a crash course for Garwin in the semi-secret technology, weapons delivery systems, and the "contest of offense vs. defense" strategic nuclear weapons, all relatively new to him.[11] The connections with Wiesner and Zacharias served as an entree to the Washington science-policy establishment, especially the White House, PSAC, and the Office of Science and Technology (OST).

It would become his lifelong milieu. It was Garwin's first real contact with "the explicit interaction of high-level scientists with the US government in attempting to influence technical programs and policy. I learned a lot," he said, "such as Zacharias's injunction 'Don't get it right, get it written.' . . . But there were some other lessons that did not sit so well. Zacharias also observed 'If I don't get the answer I want from this group, I will try again with another,' which seemed to me to conflict with Fermi's caution to avoid delusion (self or otherwise) by reporting all the data up to the stopping point of the experiment. Furthermore, while it was good technical fun reviewing technology and inventing new systems to detect and counter nuclear-armed Soviet bombers, it was perfectly clear that by the time we could have such expanded defenses, the threat would be Soviet nuclear weapons on ballistic missiles of intercontinental range—ICBMs."[12] Apocalypse wasn't likely to come on the wings of bombers.

The SAGE system was expensive, and funding was a sometimes thing. For example, for years it also was a topic of considerable debate in the defense community, partially whether it was possible for the Soviets to jam the system. Project Lamp Light concluded they could with electronic countermeasures. The results of the study were verified when the US Air Force equipped Strategic Air Command bombers with jamming equipment in a mock attack. The planes flew right in undetected.[13] Countermeasures were later developed to let the ground-based radar operators change the frequency of the radar waves to counter the Soviet methods.

Garwin said one of the things he learned at Lamp Light was the tendency of bureaucrats to emphasize the parts of a report that benefitted them, rather than give a more balanced view. For instance, he learned there was never any serious consideration to turning over America's deterrent threat to bombers on one-way missions instead of missiles. It would have been cheaper and easier to arrange. There was the possibility the bombers could be refueled and return to bases in the US after attacking Russia, although he called that a "fiction." The effect, however, was the US military never considered that the Russians could do the same thing—launch bombers on potentially suicidal missions.[14]

The newly established Office of Naval Research (ONR) had a particular interest in what came out of Lamp Light. The US Navy wanted to make sure the fleet would play a role in the SAGE expansion, and pushed the study in that direction. What did come out of Lamp Light, which described a system in which transistorized general purpose computers (an invention less than ten years old) would link all the ships into one giant computer network, spot the targets, and fire when ready. It was the beginning of the digitalization of the navy.[15]

By 1954, Garwin himself was becoming well known in Washington, DC. The Science Advisory Committee (SAC), created by Harry S. Truman in 1951, wasn't then very influential. In 1953, Dwight Eisenhower asked the committee to assess the capability of the Soviet Union to harm the US. Garwin was appointed. James Killian, then president of MIT, was selected to head the Technological Capabilities Panel.

Eisenhower considered his one great failure as president was his inability to limit nuclear weapons—everyone's including America's and Russia's. When he took office in 1953, the US had 1,200 nuclear weapons. By the time he left office in 1961, there were 22,229. According to historians, Eisenhower was hampered by domestic politics and Soviet intransigence, but he also failed to come up with a cogent plan that the Soviets were likely to accept. The men who succeeded Josef Stalin were ideologically entranced by the expansion of Communism and the belief in Communism's inevitable victory over capitalism. Arms control was not on their agenda. They were not alone. The Eisenhower administration kept building weapons and talking about limitations and the threats to humanity the weapons presented—and was sincere in the desire to reduce the threat—but simply could not get its act together. The two powers danced without touching. Who is responsible for what is still debated by historians.[16]

Back in 1951, Major General Gordon P. Saville of the US Air Force expanded Project Lincoln, a study of existing air defense made at the Lincoln Laboratory at MIT, to see how well reconnaissance could be used against a possible attack—forming the so-called Beacon Hill Study Group. They looked at how well high-altitude surveillance would work from aircraft or even balloons, and new technologies such as microwaves and infrared cameras to monitor what was happening on the ground.

In 1953, Eisenhower had formed the Technological Capabilities Panel of the Science Advisory Committee, led by Killian, to study the technology of surveillance. It worked for twenty weeks, held 307 meetings, engaged forty scientists and military experts, and in February 1955 issued its report, "Meeting the Threat of Surprise Attack." The report "deeply frightened him," Garwin said of Eisenhower.[17]

The committee concluded that while the US still had an offensive advantage (it had many more atomic bombs), it was nonetheless vulnerable to surprise attack, its mix of nuclear weapons was not diverse enough, there was no early warning system, which made the US particularly vulnerable, and the Soviets were busy building capability for long-range delivery of nuclear weapons. If it made anyone feel better, the report concluded that if a war broke out, the US "could mount a sustained air *offensive* that would inflict massive damage and would probably be conclusive in a general war." In other words, the US would be the last one standing amid the rubble of civilization.[18] Garwin was not a member of the committee, but by this time he knew everyone on the committee and he would be drawn in. One of the questions asked in the study, completed in March 1953, was whether a Soviet nuclear attack could kill fifteen million Americans in the Boston-New York megalopolis.

The committee recommended developing the Polaris, a submarine-based missile, and the DEW line across the Arctic. Edwin Land, who would later invent instant cameras at Polaroid, called for the CIA to develop high-flying reconnaissance aircraft.[19]

Eisenhower never believed the Soviets would actually launch a surprise attack on America, but the risk of being wrong was immense, the fear palpable. After all, the Japanese surprise attack on Pearl Harbor was only twelve years earlier. The North Korean invasion of South Korea was an unhappy surprise fewer than three years before, and was a major intelligence failure. Since August 29, 1949, the day the US detected the first Soviet atomic bomb test, the threat of a nuclear Pearl Harbor loomed. The Soviets posed a dystopian danger.[20]

The Soviets had themselves some genuine concerns. While the Western countries were maritime nations, protected by oceans and sea power, the Soviet Union was not and, a dozen years earlier, had been the victims of a massive land attack by Germany that killed tens of millions of its people.[21]

Russia's technology in bombers and missiles frightened Washington. Eisenhower was particularly fearful of the bombers; most of the military experts were more afraid of long-range guided missiles. Attempts in the year before by the United Nations to come up with a disarmament plan failed utterly. The technological hurdle that would haunt all disarmament negotiations was verification. What means of inspection can you employ that would assure either side that the other wasn't planning an immediate attack? What kind of surveillance would be acceptable for each to have its territory watched by a potential enemy? There was very little human intelligence—spies—behind what Churchill called the "Iron Curtain." Soviet counter-intelligence was excellent and Western spies few. The Soviets had an almost pathological reliance on secrecy, and any transparency introduced for the purpose of arms control would be a tough sell. Many experts said their obsession for secrecy was an advantage over the US. And there was geography: Russia had the largest landmass in the world, and surveillance would be no easy matter.[22]

Opening Russia to inspection would be a great advantage to the US, but how do you do that? It worked the other way too: opening US airspace to the Soviets was problematic on the American side and would be advantageous to the Soviets. Nonetheless, Eisenhower was sure the answer was in technology, and the discussions over the possibility of sudden attacks became the place where technology crashed into politics.

The summer of 1958, the National Security Council issued a classified report urging adoption of a satellite surveillance system, afraid that high-flying U-2 reconnaissance aircraft were no longer a viable solution. (In 1960, a U-2 piloted by Francis Gary Power would be shot down.) Another problem was that America's allies disagreed on many issues, which limited possibilities. Not every country agreed that transparency of its own military installations was a good thing while everyone in the West agreed the Soviets should permit it. The Russians were a unified bloc, "the Soviet Union and four stooges," Garwin said.[23] Russia's allies were in no danger of being independent; what Moscow wanted was what the Warsaw Pact wanted, a unified position. America, on the other hand, did not give orders to its allies, and one in particular, France's Charles de Gaulle, was not interested in playing well with his English-speaking allies.

In January 1958, Eisenhower wrote a letter to then premier Nikola Bulganin suggesting a conference to discuss ways of avoiding surprise attacks in North America, Europe, or the Soviet Union. Bulganin agreed, and a meeting was called by the United Nations in Geneva in November of that year. Five countries would attend for each side. For the West: US, Great Britain, France, Canada, and Italy. For the East: the Soviet Union, Poland, Czechoslovakia, Romania, and Albania, although in reality all of it was the Soviet Union. A month later, Bulganin was forced out of office.

Nikita Khrushchev, who replaced him as premier, agreed on July 2. The next day, Secretary of State John Foster Dulles asked Killian, then special assistant for science and technology, to look into the scientific and technical problems with early warning abilities and capabilities. What should the US be looking for to spot a buildup for a planned surprise attack? How would America reduce the possibility of an accidental attack? "What would be the most important objects and means of inspection and control?"

Killian responded that it couldn't be done, unless it was linked to reduction in weapons and delivery systems, in other words linking the technological problems to political issues—exactly what the Soviets were saying. There simply were too many nuclear weapons around to monitor. A working group at the White House, headed by George Kistiakowsky, agreed with Killian's position.[24] Eisenhower wanted a technological solution. Everyone else seemed to realize it wouldn't happen.

It would be the sticking point that doomed the conference before it even began. Technology alone wouldn't solve the problem.

Eisenhower asked Kistiakowsky, General Otto Weyland, and William J. Porter, a diplomat, to form a group to go to Geneva for the surprise attack meeting. The three assembled a force of one hundred scientists, military officers, and strategy specialists, including Garwin representing PSAC, to attend the meeting. Curiously, few experts came from the State Department, reflecting Eisenhower's belief that this was a technological issue, not a diplomatic one, and they did not have anything like the expertise needed to negotiate on this matter. The American committee also brought with them to Geneva a wide breadth of recommendations, such as monitoring Soviet supply lines from the factories to the weapons site, ideas the Soviets had no thought of permitting even if it was possible. It would require some 35,000 people. To protect the Arctic, the Americans proposed to put two squadrons of RB-47 bombers, one in Alaska, the other in Greenland, with more than 700 people assigned to each.[25] Presumably, the Soviets would be permitted a similar buildup.

In reality, the US contingent was not fully competent to negotiate.[26]

The group had met for several weeks in Washington for briefings by the intelligence agencies and the military. Wiesner served as technical staff director. The briefings were on the dangers of surprise attacks, "what did we know, how surprising it would be, what in principle we could do to prevent it," Garwin said.[27]

In mid-November, the group traveled to the United Nations facility by the lake on the eastern outskirts of Geneva. The two sides would meet during the day, then gather among themselves later to discuss how to respond to what had happened that day. The American contingent worked at offices at the Hotel du Rhone downtown. One floor up was another delegation of Americans who were attending an international conference—a Conference of Experts—on detecting what nuclear tests would be necessary if there ever was to be a comprehensive nuclear test ban treaty.

Garwin brought his family, and they were living in a *pension*. They eventually moved into an apartment—with a maid. She would come back to New York with them later, and then, the next year, when Garwin went to work in Geneva, she returned to Europe.

The Russian negotiators were watched by the secret police, minders. "They were warned there would be Western spies and entrapment and all that," Garwin said. "They were told you should give a kopek to get a rubble,"[28] meaning they were to give as little information as possible and obtain as much as possible, a typical stance Garwin would encounter for decades, and that made negotiating with the Soviets particularly frustrating. One of the Soviets, he remembered, was named Tsarapkin. He was known as "Scratchy Tsarapkin," because he was somewhat "abrupt."

"They didn't want to get anywhere," Garwin said.[29] There were all kinds of things they could have agreed on, but they were not interested. The Soviets also were upset that the US was sending long-range bombers over Europe and the Arctic loaded with nuclear weapons, "a dangerous practice," and "a threat to peace," the Soviets said in a letter to the UN.[30] The US said they were only training missions; the Soviets pointed out you didn't need to train with aircraft actually carrying nuclear weapons. They called on the UN to ban the flights—unsuccessfully.

"The Soviet Government cannot aid and abet those whose desire is not to avert the danger of surprise attack but to collect intelligence information about the latest atomic, hydrogen, rocket and other weapons of the Soviet Union," Moscow explained.[31]

Since the American delegation did not have all its arguments in place, the talks were doomed from the beginning. "After six weeks we agreed on the title of the conference," Garwin said, sarcastically: "The Conference of Experts for the Study of Possible Measure Which Might Be Helpful in Preventing Surprise Attack."[32]

"There was really not much I could do for the surprise attack people," he said, "so I kibitzed with the [detection] people and did some work for them upstairs. In Surprise Attack, we decided early on there was not much we could do to eliminating attacks by negotiation but that we could certainly do a lot to take the surprise out of it, to allow preparation for retaliation or defense."[33] In other words, there was no way to stop an attack, but there might be a few ways of giving a warning that one was coming so retaliation was possible.

During the two months of talks, there were heavy bombers taking off from Soviet and American airfields with nuclear weapons. Missiles sat brooding in silos tipped with atomic or hydrogen warheads. By now, both sides had hydrogen bombs. There would be enormous destruction in store if things got really bad.

How would you eliminate surprise? One method considered was posting people on the ground at airfields and missile silos watching for planes taking off or missile launches. If they noticed something, they would somehow communicate back home to give a warning.

"You could have a little kiosk, with air-conditioning and heating and poison gas injectable on demand to be realistic," Garwin joked. "They couldn't send home a message saying a surprise attack was on the way because they would be dead long before and their communication system would be interrupted."[34] One idea that would work would have an instant communications system, even from the wilds of Siberia, that would send one message when nothing was happening and perhaps send a different message if something was. The Russians did not have a reliable communications system of their own and probably would have liked to acquire one, so the idea of building such a system might seem attractive, he thought.

"My role was pretty big," Garwin said. "It was to design a communications system in case we couldn't rely on the Soviet [system] ... and to identify the Soviet missile launching and airplane launching sites."[35] Garwin designed geosynchronous satellites that would receive Teletype-rate signals, a few hundred characters per minute, and would be generated automatically by the missile-watcher equipment with the operator holding down a "deadman's key." The all-is-quiet message would go through as long as he or she held the switch. If the planes took off or the missiles roared out of their silos, the missile watcher would just let go of the key and the transmission would cease, alerting the other side that an attack was on the way. Killing the watchman would have the same result. His foot would come off the switch. Messages would be encrypted so that the Soviets couldn't send out one saying all was well as the planes took off.

Then there was the idea of aerial reconnaissance.

The problem with aerial reconnaissance had to do with clouds—you couldn't see through them. Garwin decided to do a little operations research. If you had one hundred airfields, how many times would you have to send reconnaissance missions over the area to guarantee seeing the airfields and what was happening on them 99.9 percent of the time? It wouldn't do to have the Russians wait for a cloudy day to launch an attack. (Under the Open Skies Treaty with Russia, Russian military now fly with American planes over America, and American military fly on Russian planes over Russia.)

The math involved was onerous, so, with little else to do at his conference, Garwin decided to go to CERN, the European nuclear research center on the Swiss-French border, since they had a computer—albeit a primitive one—called a Ferrenti Mercury that might work out the probabilities. The trip was also one of pleasure. Garwin had no official business, but he walked across the lawn, opened a door into a lab, and introduced himself to the first person he met. The physicists at CERN knew who he was because they were using some of the equipment he had invented at Chicago earlier in the decade. He was that well known in the physics community by then. So he was booked for time on the Mercury. But the numbers involved in the equations proved to be too big even for the computer. He worked around the limitations by using an approximation. While he was there—and since the surprise attack conference seemed to be going nowhere—he plugged some equations into the machine for the nuclear test conference, which was going on simultaneously, although officially he had nothing to do with them. The problem he worked on was detecting signals of underground tests with seismometers.

People had different calculations of how large a signal there would be at a seismometer at a given distance from an explosion, Garwin said. Both sides, the Soviet Union and the West, did agree on a standard seismometer. A great deal of work had been done to understand the limitations of detection systems—it was complicated. Can you triangulate seismometers to locate an explosion? How do you filter out local noises such as trees shaking the ground in the wind, or more mere ambient throbbing from nearby surf? And how do you distinguish an earthquake from a nuclear test? What if they happen simultaneously? Furthermore, if you need to monitor a test area you first need to know where it is. If someone drills a borehole to test a nuke, how deep does that hole have to go before it is invisible to detectors? A bomber can drop a bomb anyplace, and a ship can drop one anywhere in the broad ocean. How do you detect those?

Garwin would visit CERN regularly to talk physics, and the

people he met there would become some of his closest friends. He and his family went for a year's sabbatical the following year.

The test negotiators had run into a translation problem about the sensitivity of the seismometer that would be used to measure any explosions, and had come up with different numbers. Garwin had straightened out the situation at the CERN computer and was called to support what he found. Lois had been allowed into the meeting, and in a rare moment, she saw him at work. He gave a presentation, was challenged by the Soviets, and responded.

The surprise attack conference continued to go nowhere, but Garwin said both sides were civil to one another and got to know the other side. Some of them, he said, "were quite good,"[36] and the familiarity would pay off in other conferences.

Christmas rolled around, and the Western nations wanted a break for the holidays, but the Russians were not interested, Christmas being less important to them. The West refused to continue negotiations and wouldn't agree to resuming on January 5 despite prodding from the Soviets. They never met again.[37]

The detection meeting was a rehearsal for decades of arguments and eventually morphed into the 1963 Limited Test Ban Treaty.

Each government had different views about disarmament, Garwin said, and different people in each government had varied views.[38] Many in the United States wanted more weapons, not fewer.

"That's why you have a president and leadership," he said, "so the President can at least get these things studied."[39] Of course, that doesn't end it. The military and the contractors go to Congress formally or informally, and the best-laid plans go astray regardless of what the president wanted. The White House was relying more on these weapons and building while at the same time wanting to reduce them.

The conference of experts provided evidence that seemed to satisfy most that nuclear explosions could be detected in the atmosphere and in the oceans and with satellites, in space. It laid the grounds for the 1963 Limited Nuclear Test Ban Treaty, dealing with everything except underground explosions. It was limited, Garwin said, because the US insisted on ten onsite inspections, and the Soviets would allow fewer. Both sides could have compromised on five if they really wanted a test ban. The holdup was at least the fault of the British, who didn't want to pay for the inspections. Garwin suggested they find a philanthropist to pay for them.[40] Behind that excuse, the British felt they were being left behind in the development of the weapons and wanted to catch up, but that was not a socially acceptable excuse.

The surprise attack conference was the last United Nations conference Garwin would attend. He returned to the US and IBM in December, and in August 1958 used a Ford Foundation grant to return to CERN.

CERN, or Conseil Européen pour la Recherche Nucléaire (the European Organization for Nuclear Research), was founded in 1954, one of the first European scientific collaborations. Now one of the largest research facilities in the world, it was just beginning when Garwin visited. It is near Geneva, and part of its accelerator is in France, part in Switzerland with labs and offices straddling both countries.

The principle—and it is true of all accelerators—is that if you crash atoms and parts of atoms together at very high speed and sufficient energy, things fall out. Sensitive detectors spot those things and measure them. The things could only live for a millionth of a second (or a billionth of a millionth), but it is the best way to define what's inside an atom, what matter is made of.

Leon Lederman had been running a series of experiments on the muon at CERN, but he had gone back to the States. Garwin was not going out to do muon work, but because of all the time he had spent at the various conferences, he wanted to spend an idyllic time with his family in Europe and sit in CERN's growing library to catch up on physics. Being Garwin, that's not what happened.

He packed his family in New York, along with Marie, the maid he had brought back from Geneva, and boarded the French ocean liner *Liberté*, not the most luxurious liner traveling the Atlantic but "pretty fancy," he said. They packed two steamer trunks ("Yes," he said, "they really did exist!") including one loaded with toilet paper and peanut butter. The toilet paper was to make up for the coarse paper used in Europe then, the peanut butter because Tom was a picky eater and peanut butter and jelly sandwiches were a mainstay of his diet. You could always get good preserves in Europe, but Europeans knew nothing of peanut butter.[41]

At Le Havre, they loaded a rental car and drove leisurely toward Switzerland, seeing the countryside, stopping at good restaurants, having a vacation. "It was," Garwin said, "a good time."

At Geneva, the family moved into an apartment they inherited from Lederman on the Rhone River, close to bus and streetcar lines. The two boys were enrolled in the international school; Laura, then only two, stayed at home with Lois and Marie.

Lederman had organized a group of young physicists to continue their research into the muon. The goal was to find out if the muon, born in cosmic rays, was a Dirac particle, and Lederman's and Garwin's earlier work indicated that they were. Protons and electrons are Dirac particles, named for the British mathematician Paul Dirac who described them in a famous paper in 1921. His equation—which some consider the most beautiful in all science even if it is incomprehensible to non-scientists—describes particles straddling the worlds of relativity and quantum mechanics. (Dirac also predicted the existence of antimatter.) Lederman's troops were trying to find the magnetic moment of the muon. CERN apparently had instruments delicate and sensitive enough to find out.

They were stumped. Someone in the group decided that if they didn't have Lederman to lead them, they had Richard Garwin. Garwin refused. Doing experiments was not why he was in Geneva. One of the young scientists was the son-in-law of CERN's director of research, Gilberto Bernardini, and he went to Garwin and begged him to take over. Garwin still refused until Bernardini came to him "crying."[42] He relented but on his terms: the team would not be run as a democracy. Anyone can make a suggestion on how to proceed, Garwin would decide what to do, and make assignments. The team would work "American style," which meant sometimes weekends. It also meant working through the December holidays.

They agreed on the timetable, except for the holidays. That, they refused. CERN shut down for the holidays. "That was a bridge too far," Garwin said.

The team was exceptional. One member, Georges Charpak, went on to win a Nobel Prize in 1992. Another, Francis Farley, spent the rest of his career expanding on the CERN experiment.

It took two years to get the experiment done right (the last part with Garwin back in the States talking to the researchers by Teletype and telephone).

Meanwhile, the Garwins saw Europe, driving down from Geneva, heading south toward the Mediterranean, and then along the coastal highways, past Livorno, to Naples, Rome, and Pompeii, later north to Florence and Venice.

Weekends were for skiing. For a while, at least, Garwin did not have to interact with the Washington-scientific complex.

When the Ford grant expired, they flew home (without Marie), back to IBM and more muon work at Columbia—and finally back to government work.

Garwin was not involved in Project Starfish Prime; he just had to bail the Atomic Energy Commission (AEC) out of the mess it created.

In July 9, 1962, the DOE fired a missile into space from Johnston Island in the Pacific. It carried a hydrogen bomb, 1.4 megatons, one hundred times the explosive power of the Hiroshima bomb. It exploded 400 kilometers (248 miles) above the Earth and above the Earth's atmosphere over a point nineteen miles from Johnston Island. The Russians sent ships to the area to watch.

The test was largely the influence of two brother scientists at RAND, Dick and Al Latter, disciples of Edward Teller. According to Wolfgang "Pief" Panofsky, who was an ally of Garwin in the disarmament battles, the Latters spent their time figuring out ways the Soviet Union could violate testing bans, largely by inventing procedures for them.[43]

At the time, a space explosion violated no treaties. The plan was that the radioactive fission products would remain at the point of explosion and the electrons from the decay would follow the magnetic lines around our planet and fall harmlessly back to Earth, and that would be the end of it. The worst that would happen, supporters felt, was that radio astronomers would have to put up with a good bit of noise in their readings for a few weeks. The Russian cosmonauts then in space were, it was assumed, safe. It was an easy assumption to make, and Garwin admitted he would not have disagreed.[44]

Not everyone was pleased or sanguine about the experiment. Scientists in some of the government labs were worried about what would happen if a nuclear bomb went off that close to the Earth in what is called low Earth orbit. The test did not seem to have any military value. There was also the real danger of knocking out satellites with the electromagnetic discharge. (The blast did, in fact, kill seven satellites, including Telstar, the first communications satellite). More profoundly, President Kennedy had already announced the program to land humans on the moon—the Apollo program. Would the astronauts have to fly through a sea of radiation to go to the moon? Would it make the mission impossible? More dramatically, should the US tell the Russians to get their cosmonauts down?[45]

The impact of the blast exceeded all predictions, and things fell apart quickly. It could be seen in the sky from Hawaii, where it knocked out streetlamps, set off burglar alarms, and blacked out a microwave relay station with its electromagnetic pulse (EMP). The Royal New Zealand Air Force used the light of the Starfish blast for a submarine hunting exercise. Many of the instruments measuring the blast went off scale.

Radiation was trapped in the Van Allen belts, a belt of radiation that circles the Earth, and it did not look like it was going away soon. In fact, it lasted five years. The Earth's magnetic field held the high-energy electrons from the blast, creating what Jack Ruina, then head of the Advanced Research Projects Agency (ARPA), said was "an x-ray machine that went from the North Pole to the South Pole."[46]

Wolfgang "Pief" Panofsky, who headed the Strategic Military Panel, was on vacation in Baja California in Mexico, so Wiesner, Kennedy's science advisor, asked Garwin to take over the government's reaction.

"I needed to get up to speed on what was happening, both on the nuclear weapon aspects of it and the trapped electron aspects of it that I had never studied in detail before. And I needed to contact the various players in this," Garwin told Dan Ford. The tests were a deep secret. Even people like Garwin, who had the highest possible security, and Panofsky did not know about it. It involved potential weapons, which accounted for some of the secrecy, and it was in some ways a dingbat idea. "I had this very intense and sudden immersion— full-time for two weeks, as I recall—in what was possibly a serious, national emergency and met a lot of people in fields with which I was not familiar."[47]

The idea of an EMP from nuclear explosions wasn't entirely new territory for Garwin. He'd written a paper on it in 1954—still classified—involving the form of the explosion that sets the EMP flying outward. The explosion of the nuclear device would be spherical, outward, Garwin said, but the generation of the early electromagnetic waves depends on electrons moving across the line of sight, and the actual mechanism responsible for the intense electromagnetic pulse was missed by all until after the Starfish Prime test.[48]

Wiesner set up an appointment with Kennedy at the White House to tell the president how worried he should be. Present also were Garwin, Glenn Seaborg, then head of the Atomic Energy Commission, and Carl Kayser, the deputy national security advisor.

"What's happening? What have we done?" Kennedy asked.[49]

Garwin explained that the Van Allen belts are streams of electrons that ride the lines of the Earth's magnetic shield. One of the purposes of Starfish was to see if those electrons could be used to interfere with Soviet communications by pumping fission-product electron volts in the belts. The military also wanted to know how long they lasted. Several earlier tests had shown promise.

Kennedy asked how long the electrons would last, and Garwin had to say he didn't know. He could estimate.

"So I told him, 'Well, to within an order of magnitude, it was this or that.'" Kennedy had never heard the phrase "order of magnitude," apparently, and Garwin had to explain it meant a factor of ten.[50]

Kennedy turned to Seaborg and joked, "Now Glenn when you tell me something, I should believe it only to an order of magnitude."[51]

Garwin thought word of what had happened was sure to get out and embarrass the government, but fortuitously, the *New York Times* and all the newspapers in New York were in the middle of a typographer's union strike, so the crisis did not get extreme.

Garwin devised a method of sweeping the electrons out of the way, using uranium or lead foils, but that proved unnecessary. The planners had missed a point: "The 400 kilometers is low enough so the Van Allen belts don't come down that low, and if you have fission products decay then those electrons would stay in the earth's magnetic field. . . . The people planning the experiment missed the idea [that] the nuclear explosion would create an enormously conducting plasma that would push aside the earth's magnetic field. But the earth's magnetic field is diverging upward toward the magnetic poles. The magnetic bubble likes to be in a region of small magnetic fields, a higher altitude," Garwin said. The bubble bubbled the electrons up out of the atmosphere.[52]

The cosmonauts' radiation badges showed no sign of dangerous doses, Garwin said. Starfish reminded him of Edward Teller's fear that Trinity would ignite the atmosphere.[53]

One of the reasons Garwin took the job at IBM was his belief that he was, at heart, an inventor. "I could do useful things that IBM could develop and sell," he said.[54] His first project was data storage for computers, which until the last decade, was very expensive. Storage in the 1950s was something like $20 a bit of information. He worked on several ideas that did not pan out, but the problem interested him. He moved to another lab in an old Columbia fraternity house on West

115th Street, where IBM engineers were working on storage as well, using nuclear magnetic resonance (NMR), the same technology now used in medicine as MRI. The researchers used hair oil, an emulsion of water in oil, in the magnet. The hair oil made it possible to store thousands of the magnetic information pulses in the water droplets. It was overtaken by other technologies.

In 1965–66 he was promoted to director of applied research at IBM. He had some 600 people in three labs, including one in San Jose. A chemist there had been assigned to propose a plain-paper copier to compete with Xerox, something that would use different technology so IBM didn't have to license Xerox's patents. The goal was to make a machine that could reproduce several hundred thousand copies before it had to be serviced. Garwin chose among various technologies one that used an organic photoconductor film a bit like Saran Wrap wrapped around a metal drum instead of a metal-coated drum to carry the image.

"It was a considerable success," Garwin said of the copier.[55]

"When it came to replacing mechanical ink-on-paper printers, there were a hundred different technical options, and they had all sorts of people working on it using various technologies. And [Garwin's] job was to say, 'Check that one.' Option 13C. Well, option 13C was the one that worked. He has a very long list of patents that they profited from," Dan Ford said.[56]

Garwin has fifty patents.

Then IBM wanted a nonimpact printer. Management favored an approach using a cathode ray tube and optical fibers. Garwin didn't think that was the best technology, but his suggestion, using a rotating mirror deflecting a laser beam, was rejected. A year later they admitted that Garwin's system would have been the best choice. Garwin's engineering assistant from Watson lab had moved to IBM in San Jose and had kept a rotating mirror-operated device under his desk from the previous year. Garwin told management to take a look at it.[57] It was adopted and became the IBM 3800 printer, the first laser printer, 240 pages a minute on wide green-and-white striped and perforated paper.

Another project was the gaze-controlled computer that fit on the helmet of a pilot for a virtual cockpit, or that could be used to work a computer. The computer watched the operator's eye motion, where he or she looked on the screen. The original design was for a virtual cockpit, one without switches or gauges. The defense electronic outfits weren't in the least bit interested. IBM wasn't either. Now engineers are designing the "cockpit of the 21st century." Garwin and his colleague Jim Levine had invented the cockpit of the 1960s and demonstrated some of the technology, but no one cared.

Garwin had the problem of selling his ideas in the company, even as an enlightened one as IBM at the time. He said,

> The job of the innovator is not done when he or she thinks of a new way of doing something. We've done exactly the same thing, as I pointed out, in the laser printer. I spoke for years with the people in the office products division about making the IBM electrophotographic copier into a laser printer, and there were all kinds of arguments why they couldn't do it. The market wasn't there and so on. Well, the proper market of course is for a computer output printer, not an office product printer, and there the reason good people didn't pick it up is that they were in the wrong division. If they had made this thing it wouldn't have been used in the computer division, and the computer folks when I talked to them about it said the copier didn't have the reliability and longevity required of a computer room printer. But that's only a matter of engineering, and I could have done that. There was no reason to discard it from that point of view.
>
> Later when I was proposing the laser rotating mirror for the electro optical transducer in the IBM 3800, the ones in the proper division didn't want it because they had their own solution, they thought. So they looked for not any kind of balanced assessment, but arguments against it and in favor of theirs. Many people don't even know they're doing this.[58]

Garwin and Levine were also ahead of the world with another invention, the touch screen, and again had problems getting others to understand what it could do. Levine, Garwin, and an engineer, Mike Schappert, demonstrated a touch screen for IBM PCs in the 1970s and '80s. They were able to demonstrate the technology on color monitors, including a way to draw on the screen using color styluses, and Levine showed a ticketing kiosk that could have been turned into an ATM. IBM seemed peculiarly resistant.

"IBM wasn't ready for it," Garwin said. "Instead, the company put a little red button in the middle of their keyboard [of their laptops] to serve the same function as the touch screen."[59] The company did turn out touch screen products, including touch screen overlays for the IBM Industrial PCs. IBM also used to produce lecterns, the InfoWindow, which was a way for lecturers to show illustrations, a hardware version of what would become something like PowerPoint. The lecturer could call up "slides" on a seventeen-inch plasma display and annotate them with a finger or stylus and have the composite displayed instantaneously on a large digital color projection screen.

One IBM manager was particularly obdurate about the technology even when shown a model of an ATM. IBM made ATMs. People wouldn't want to touch the screen because other people have touched it, he reasoned. Garwin responded by asking about the keyboard. People touch that. The manager replied yes, but when the transaction was done the screen would go black and people could see the smudges. "Okay," Garwin responded, "don't go to black. Go to gray."[60]

IBM missed the boat. Garwin still wonders what would have happened if he and Levine had gone into business with that technology. He mentions it often.

"Along the way, we invented gesture recognition to supplant the buttons on a mouse, when one is using a touch screen. But it was not the commercial success it deserved to be—because we didn't have the fire in the belly either to go it alone in the hopes of getting rich and famous, or the internal persuasion of people who were much more monomaniacal than we."[61]

It wasn't until the advent of the iPhone that the value of touch screens became obvious.

Garwin also invented protection devices for the hard drives of laptops. If you drop your laptop and it still works, you have him to thank, and while at the Military Aircraft Panel for PSAC, he contributed to the development of GPS satellites, a technology that needed to take into account Einstein's theory of general relativity.[62]

As was often the case, he was ahead of the people who made decisions. "In 1980, I had pressed the head of the uniformed Air Force, General Lew Allen, to move ahead with GPS; instead, he reprogrammed the entire $2 million GPS allocation to another purpose because he said he had higher priorities. He knew Harold Brown, the secretary of defense, would put the money back."[63] At Garwin's urging, he did.

"So I got the money," Garwin said, and years later, Allen admitted he might have listened better.[64]

One of his most famous and at the same time quietest accomplishments—quiet because it was of interest mostly to scientists and engineers and because it is difficult to describe in words—is his effort on behalf of the "Fast Fourier Transform" (FFT). This is an algorithm that can be used for finding patterns in signals over time, but also in the efficient transmission of pictures or television and to compress large computer files such as images. It was used in the American computer in Thailand that helped find the North Vietnamese on the Ho Chi Minh Trail crossing the McNamara line.[65] It helps to detect planets circling other suns in vast amounts of astronomic data and is involved in every aspect of electronics and communications.

Jean-Baptiste Joseph Fourier was a physicist and mathematician in early nineteenth-century France. He originated the highly counterintuitive Fourier Transform for the description of heat transport in solids. The best analogy to explain it comes from music. Say someone plays the middle C on a piano. The piano sends out waves of air molecules at the frequency of 261 vibrations a second. But your ears don't hear it as waves. They hear middle C. Middle C is the peak of the Fourier Transform of the vibrating air molecules.

Garwin did not discover the Fast Fourier Transform but is credited with bringing it into the mainstream of science and engineering. The reason is simple: the ordinary Fourier Transform on a project involving 1,000 data points required the order of a million multiplications and additions the old way, while the FFT requires only 7,000 for the same number of points. A more current problem would be to apply corrections to a high-definition picture, which has about 12.7 million pixels. Assuming a processor that can do three calculations a nanosecond, a traditional FT would take 54,000 seconds—15 hours to compute. Using the same processing speed, a complete FFT would take a total of 0.1 seconds.

Garwin wanted to use the Fourier Transform for determining the structure of solid helium-3, on which he was doing magnetic resonance experiments in the early 1960s. He created a computer program to model the interaction between adjacent helium-3 atoms and found that at the lowest temperature, the pattern of nuclear spins "relaxed" to some low-energy state. But what was the pattern of spin orientation in the crystal?

At one of the monthly meetings of Kennedy's PSAC, Garwin sat next to John Tukey, a statistician of Princeton University and Bell Labs, at the end of a huge table in a conference room, and noticed that Tukey was writing Fourier series on a pad of paper with his left hand, a sort of a doodle. Garwin inquired whether Tukey knew something he didn't know, and Tukey said he was working on a method that could double the number of points in a Fourier series in less than double the time.

Garwin, immediately on his return to IBM, asked that the head of mathematics in the IBM Research Division designate a numerical analyst to work with Tukey to code Tukey's method into a computer. J. W. Cooley was the man, and he was ordered to Princeton. He coded the FFT algorithm in the scientific computer language, Fortran. It is now known as the Cooley-Tukey FFT algorithm.[66]

In fact, many mathematicians had found methods to reduce the number of calculations required to complete a FT, a fact that came out as the FFT became more popularized and made its way into other scientific disciplines.

That other people had discovered and published similar algorithms previously shows the significance of Garwin's role. He wrote dozens of letters to other scientists and heads of laboratories, and when he visited other labs he made sure to mention what he had learned. Garwin was able to see the real power of the FFT and the broad applications outside of individual calculations, and the FFT is now universal thanks to his proselytizing.

The development of FFT was really something like a revolution. When the method was first created, it was thought of as a convenience for people doing work that involved Fourier Transforms and who wanted to spend less time and computing power on each problem. However with the significant improvements in time and computational power, suddenly FFT could be applied to problems that could not have been analyzed by the traditional methods because of their complexity. "FFT methods now permit one to do digital processing on acoustic data in real time. One prediction is that someday radio tuners will operate with digital processing units," Garwin once predicted.[67] That day came long ago.

Cooley and Tukey published their paper in 1965. Garwin ended up using a different method for his helium experiment.

"I disclaim any intellectual contribution to the development of the Fast Fourier Transform," Garwin said. He realized that it would be valuable across science and technology. "I do know a good thing when I see one."[68]

Garwin added, "When I learned of this tremendously exciting prospect, I resolved that I would do more than use it to solve just my own problems; I would do what was necessary to make it practically available to everyone."[69]

It had commercial value too. Garwin encouraged IBM technical sales people to explain at meetings how important the FFT was, how the enormous speedup in computation was helped by the fact that the Cooley implementation used no more computer memory than the initial input data—another reason to buy IBM computers.

"I think we probably helped out a little with IBM sales teams who were attempting to sell computers to people who were doing vibration analysis of complex systems," Garwin said.[70]

He is credited with helping to make the FFT a common tool in virtually every aspect of science and technology, to the relief of graduate students who have escaped the fate of doing millions of calculations themselves.

CHAPTER SEVEN

ADVISING PRESIDENTS—OR NOT

In 1951, Harry S. Truman created a Science Advisory Committee (SAC) to do what the name implied. He put Oliver Buckley, retiring president of AT&T Bell Labs, an economist, in charge. Later, under Eisenhower, it created the Technological Capabilities Panel. It was, Garwin said, very secret and looked at such things as the effects of a firestorm after a nuclear attack on an American city.[1] It remained mostly in the shadows of government until October 4, 1957, when the Soviet Union launched Sputnik and scared the hell out of the United States. Nothing had prepared the US for the shock of having the Soviets jump into space first. Nothing had prepared the US for the idea that the Soviets were at least technologically equals. This was the height of the Cold War, a few years after the Korean War, and nerves were tight and sensitive. The popular joke went that the Soviets could never sneak an atomic bomb into an American city in a suitcase because they had not yet perfected the technology of the suitcase. That stopped being funny. Sputnik wasn't actually the issue, although Americans could track its beeps across the sky, the electronic version of a Bronx cheer. The issue was that the rocket that carried it into orbit was designed to deliver hydrogen bombs to American cities, and clearly they had that technology down and the US did not. Missiles carrying thermonuclear warheads were the universal nightmare of the time.

President Dwight Eisenhower's response was to make sure the president, whoever he was, had the best scientific advice possible, something that was inconsistent before. "He understood the military and didn't trust them one bit," Garwin said. "He decided he needed scientific expertise, particularly with the confrontation with the

Soviet Union, to defend the country against planes carrying nuclear weapons, chemical and biological warfare."[2] Eisenhower named James Killian as his special assistant for science and technology, and converted the moribund SAC of the Office of Defense Mobilization into the President's Science Advisory Committee (PSAC). He had asked them to undertake a "technological capabilities" study. Killian, whose degree was in management, was a master at playing bureaucracies, and his presence as one of the most prominent public scientists in the country assured PSAC clout around town.

Many of the members were drawn from the ranks of the Manhattan Project. In many respects the work they did on nuclear weapons was one of the reasons they joined. Many felt science, which had created the weapons, had a responsibility to control the weapons, and a high-level advisory committee, one with a direct line to the White House, was one way of doing it. They brought to government technological skepticism.

Garwin was named to the committee in 1962, becoming one of the youngest on the board. He had already served on PSAC's Strategic Weapons Panel—the "weapons" were nuclear and thermonuclear bombs and warheads, and the missiles to deliver or counter them. Alvin Weinberg, the physicist who ran the laboratory at Oak Ridge, Tennessee, was one of the few to oppose his nomination, thinking Garwin lacked judgment, and he pushed for Eugene Wigner instead.

Not all Americans were convinced that science had a proper role in interacting with government. The function of PSAC was to deal with the dichotomy of trying to help the government support science and to have science help the government, which is never easy. Eisenhower's passion was a ban on testing nuclear weapons on the ground.

Part of the problem, Killian felt, was Eisenhower, who had perhaps too much respect and trust in the scientists advising him. He believed they were impartial, selfless, almost to the point of naïveté. Not everyone in Washington or academe agreed. "He somehow came to have a feeling that these advisors, by virtue of being scientists, were endowed with an objectivity in technical matters that he didn't find in other advisors," Killian wrote.[3] Supporters of the advantages of totally neutral advice were countered by those who knew that Washington was first a political town and no advice was immune to political pressure. As the scientists would eventually find out, the latter were right. What would happen if the powers that be wanted something to happen and the scientists said it couldn't or shouldn't? Politicians and bureaucrats who had a vested interest in a project would not be happy. Yet many of the scientists involved in this science-government alliance believed that part of their role was advising against projects and plans. The best service they could provide would be to say no. Garwin was a specialist in that answer, making him decidedly unpopular in places around town for decades.

"I was impressed by the seriousness and dedication of these people, and their commitment to informing the President and his staff, and to providing not arguments in favor of a preconceived program, but potential solutions with their positive and negative aspects," Garwin said.[4]

It wasn't a new struggle. During the American Civil War a group of scientists, calling themselves the Scientific Lazzaroni, talked Congress and Abraham Lincoln into forming the National Academy of Sciences (NAS), the first major collaboration between science and government. The purpose was the same: help science by influencing the government and help government by giving it access to the advice and truth of science. The Royal Society of London and the French Academy in Paris had the same roles.

Woodrow Wilson created the National Research Council (NRC) within NAS to handle the actual science during World War I. Franklin Roosevelt had a Science Advisory Board (SAB) to work through NAS. He did not provide funding for two years; it was the height of the Great Depression, and some thought that science and technology were somehow at least partially to blame for the economic disaster. Karl Compton, a physicist and also an MIT president, was the chairman. One of his goals was to convince the public that science was not to blame, but at the same time he held out the possibility that science could solve some of the issues facing the country. He brought social scientists into the mix, a first. Washington's scientific claque opposed a structure in which outsiders would give advice. Eventually money meant for research went to relieving the unemployed. Research budgets felt the shock.

An entire flock of acronyms for scientific advisory boards flew through Roosevelt's New Deal, a picture complicated by the fact that the agencies kept changing their names to satisfy bureaucratic whims. The scientists were caught not only between the needs of science and government, but the boundary between science and technology, and pure versus applied science. While some of pure science leads no place in particular, often great, unexpected things come out of it. Einstein's theory of special relativity helps the GPS system in today's cars. James Clerk Maxwell never anticipated *I Love Lucy*, or television, in his equations. There is no clear boundary: pure science can lead to the application of technology and develop radar and the atomic and hydrogen bombs that helped win the war.

Vannevar Bush's Office of Scientific Research and Development (OSRD) was in charge of most of the science for the war effort during World War II, even playing a role in starting research into the atomic bomb. Little science outside of the war effort got funded, and many scientists, especially the young ones, had gone into the military. War often prods invention. The programmable electronic computer was invented by engineers in order to compute the trajectory of artillery shells. The inventors had no idea what would eventually happen to their invention, and the only reason they got funding from the US Army was because someone had to fire the cannons and plot trajectories for the guns before they were shipped overseas with firing charts. They had no idea, even in their wildest dreams, of the PC or the Mac or quantum computing. Nobel laureate Harold Urey, along with John von Neumann, Isidor Rabi, and others, formed an advisory committee to provide advice for the design of the Aberdeen Proving Grounds, where the artillery was tested, and von Neumann acted as an advisor to the engineers at Penn who were working on the computer.[5]

PSAC was not universally popular with the rest of the scientific community. Historian Zuoyue Wang described Edward Teller as a "one man anti-PSAC" because of their numerous clashes.[6] At one time Teller recommended using tactical nuclear bombs to carve out a new Panama Canal. That was soon dropped because of opposition, largely from PSAC and from an overwhelming number of nuclear scientists.

When President Kennedy was assassinated, PSAC had something of a crisis in confidence. In many ways the scientists had found a tribesman in Kennedy who appreciated what they did for a living. Lyndon Johnson was the opposite. Johnson himself was less inclined to look overseas for things of interest than he was in his domestic policies and eventually the Vietnam War he inherited from Kennedy. It was not clear if he knew much, if anything, about PSAC. Garwin wrote to Jerry Wiesner, then special assistant for science and technology to the president, suggesting he find a way to brief the new president so Johnson would be informed about all the resources he had available in fields such as civil defense, ballistic missiles, and education.[7]

Johnson responded to the overtures by bringing scientists into his Great Society program of social welfare, not a hard thing to do considering the sensibilities involved. They were sympathetic and enthusiastic. Johnson also pushed for more diversity in PSAC, and the organization brought in new members who conspicuously did not come from the banks of the Charles River and were not all physical scientists. They also brought in economists and social scientists. The lack of women and minority scientists remained intact despite Johnson's efforts. It was ever an old boy's network.

The committee consisted of eighteen members with one employee, David Beckler, who lasted throughout the existence of PSAC. The group's agenda generally came from the president's office, although PSAC sometimes did its own thing. Eighteen people could not cover all the science that needed to be covered, so PSAC was supported by panels with members drawn from outside the committee, experts from outside government. A typical PSAC panel consisted of two or three members and was almost always chaired by a PSAC member, together with the best experts in the country for the problem at hand. Members were encouraged to nominate others who might help. They were academic scientists, still mostly from the physical sciences. Scientists from contractors were forbidden with a few exceptions, Garwin from IBM being one of them. They were paid $35 a day plus expenses (hotels, meals, travel), which went up to $50 a day later. They served four-year terms, and members elected their own membership, although the White House had the final say.[8]

"It was a sort of self-perpetuating organization at that time," Murray Gell-Mann, a member, said. "I mean, if you weren't some kind of crazy critic or some extreme critic of the government or something you were just selected by the PSAC members and . . . the government would approve and appoint you. Later it became much more politicized."[9]

Garwin served two terms, rare but not unprecedented. He had, by the 1960s, earned a reputation for "technological rationality."

They met for two consecutive days every month. The panels would work on a report for six months, presenting a preliminary report to PSAC halfway through the study. The reports—some of which could take hours—were given around a large conference room at the Old Executive Office Building. Thirty people could sit around the table in rooms 206–208, and the walls were lined with chairs for more people. "It turned out that some of the briefers knew little more than the words of their presentation," Garwin said. "It was at that point that General Glenn Kent (an Air Force officer skilled in strategic analysis involved with ballistic missile defense and offense) revealed, 'You don't understand, the purpose of a military briefing is not to convey information but to fill time.'"[10]

For many years, the sessions included lunch at the White House cafeteria, but Garwin said many administration staffers were unhappy with outsiders intruding on their territory. President Nixon, to show his unhappiness with some of what PSAC was doing, eventually barred them from the cafeteria, and they had to eat elsewhere.

"Every year we would write the President in a Top Secret memo that the missile-defense system would have this or that performance, but that it could be nullified with technical countermeasures, with tactics, or it could be overwhelmed by numbers of incoming reentry vehicles. Or destroyed by a small fraction of the warheads," Garwin said.[11]

The regular reports were eventually published unless they were classified. PSAC's standing panels included one that specialized in antisubmarine warfare and one on military aircraft. The Strategic Military Panel dealt with such weapons as the Polaris, Poseidon, and Trident submarine-launched missiles, also Titan and Minuteman ICBMs, and with the ABM (antiballistic missile system proposals).

Each of the panels was known to wander off into other things.

One example: the aircraft committee, of which Garwin was a member, once studied transportation in the Northeast Corridor (Washington—Boston). It derived from a study of civil aircraft. PSAC was asked to look at the viability of using short takeoff or vertical takeoff aircraft to move traffic between cities. They eventually moved to the issue of fast trains, trucks and cars, and speed sensors in the highways, and—in the days before Waze, Google Maps, and smart-phones—whether it was feasible to give up-to-the-minute traffic information on road signs. Fast trains were discouraged by the study, Garwin said, because of issues with right-of-ways.[12] In Europe they elevated the rails, but in the US, where trains traditionally wound through dense urban areas to downtown stations, the problem would impede fast train growth. If fast trains were to be demonstrated, they would have to be bought from France or Japan, which had already done the research and development, instead of spending a fortune reinventing them in the US.

"I'm frugal," Garwin said.[13]

During the briefing to PSAC of Garwin's Air Traffic Control Panel report, William Hewlett, of Hewlett-Packard, demanded more consideration of other options than an air traffic control system based on satellites—the panel's clear favorite—and the panel ultimately relegated the details of their favorite system to an appendix, with a more extensive comparison with ground-based radar. That was the purpose of the preliminary report, to get input from PSAC members so that the report would have the endorsement of the parent committee.

Frugality had its disadvantages dealing with the Pentagon. The US Air Force one day decided it needed to build a huge plane, the CX-HLS, which would be twice the size of anything it had, so it would be able to transport the equipment for four army divisions in thirty days or one division in a week. It would be extraordinarily expensive. Garwin's Military Aircraft Panel advocated what he thought was a better idea. Instead of designing, building, and buying the aircraft, why not just cache the equipment and supplies around the world—East Coast of the US, the Cocos Islands in the Indian Ocean—and also station seven large ships with the caches, cargo ships like the roll-on, roll-off ships that had been built after World War II, cheap, simple, capacious? When the time came, the loaded ships and their contents would be sent off to wherever the equipment and supplies were needed and could get all the cargo to the troops faster, presuming they were near water. And the leading edge needs would be filled by ship-borne equipment.

"We said, if that's the job, we can do it seven times more cheaply with ships, and we designed the ships and we got them (government contractors) to run the calculations, and sure enough it was so. . . . We got what I really wanted, which was a chance to argue this before the committee with Charlie Hitch, the controller of the Defense Department, and Harold Brown, [then] secretary of the Air Force," Garwin said. The US Air Force was not interested. It wanted the plane. Brown rejected the idea, saying sometimes when you develop complex technology, serendipity strikes and interesting things fall out, an idea Garwin called "romantic." So did Hitch. He got so irate at the Fast Deployment Logistic Ship option that he stormed out of the room.[14]

"I still think airplanes are better," he said before slamming the door.[15] The air force now has the C-5 at a cost of more than $100 million each.

For years, nuclear bombs in Europe were deployed, thousands of them hanging on fighter planes flown by several nations. To prevent a German or another NATO pilot from taking off with a nuclear bomb under his wing without authorization, the US stationed a rifleman at the end of the runway. If a plane did take off unauthorized, the rifleman was to try to shoot it down—hardly a failsafe.[16] A pilot with authorization to take off would be told where to drop the bomb, but there was nothing preventing him from dropping it someplace else. He may have different priorities than the command.

The nukes also had safety problems. Each had a little propeller at the front that was used to rotate in the wind stream after the bomb was dropped by the bomber or fighter. The propeller turned a screw attached to a cable, a distance measurer, to ensure that the bomb armed itself *after* it was dropped, not *before*, which would have destroyed the plane, not the target. One day, during an inspection, PSAC panel member Harold Agnew noticed that the arming screw of one bomb on a plane in Germany had its safing wire not pushed in far enough to keep the propeller from rotating. That raised the possibility that the nuclear bomb would arm itself while still attached to the aircraft.

Agnew's solution, adopted by PSAC, was simple and extraordinarily expensive: a combination lock. The weapon could only be armed if the right combination was entered. Researchers at Los Alamos and Sandia labs produced electromagnetic locks at a cost of $20,000 per lock. Europe alone required 7,000 locks. Total cost: $140 million spent in a few months.[17] It was called the Preventive or Permissive Action Link, or PAL.

That wasn't the end of it, Garwin said. The US Navy objected. Part of the navy's image is that it works independently, priding itself on operating without communications with land—particularly the submarines—and had for a quarter of a century. It wanted to remain isolated from what happened on the land, making it a particularly potent threat. The navy did not want to wait until it got a combination over the radio to arm its weapons. The compromise was to put the code in a safe in each sub. The safe could only be opened by two officers entering the same code simultaneously on two different locks.[18]

Garwin held a one-day workshop on airport noise. PSAC also handled reports on coal mining, nuclear power, and assisted in the creation of NASA's Apollo lunar landing program. The committee recommended that the vehicles to be sent to the moon be assembled in Earth orbit, a plan devised generations earlier by science fiction writers. Without gravity or atmosphere, the vehicle that eventually went to the moon could be as large as NASA wanted it to be and in any shape. NASA, however, decided to use a lunar orbit to send the lander down, a less efficient system. PSAC nevertheless supported the decision.

In 1962, the biologist and author Rachel Carson wrote a series of articles in the *New Yorker* magazine that was expanded into a bestselling book, *Silent Spring,* which became perhaps one of the most influential books in history. PSAC created a panel on insecticides and pesticides, Carson's topic, especially on the use of DDT after Garwin reproduced copies of the *New Yorker* articles and brought them to a meeting. The chemical industry, and the government agencies involved with that industry, savaged the book, with some relegating the issue of DDT to merely a public relations problem.

The PSAC panel, however, took Carson seriously, though not unanimously. Garwin remembers Zacharias saying, "We're smarter than the bugs, at least until they learn to spray back."[19]

PSAC had done some environmental work before. One example was the Great Cranberry Crisis of 1959, when small amounts of an herbicide were found in the nation's cranberry crop just before Thanksgiving. Shoppers decided on a cranberry-less feast, and farmers and their political supporters were furious. The Food and Drug Administration resisted any attempt to modify the laws forbidding carcinogenic substances in food, and a PSAC panel, asked to intervene, came up with a compromise. The compromise was never adopted, but the role of PSAC in environmental matters was established, so when the report "Use of Pesticides," triggered by *Silent Spring,* came out, no one questioned why experts, famous for designing antimissile devices, should get involved with cranberries. Carson herself had used the analogy of nuclear fallout with the poison of pesticides in her book, and no one argued that Garwin and PSAC were not experts on fallout.

A preliminary draft of the report was hugely controversial in Washington, with the Department of Agriculture leading the charge against it. The final report was revised somewhat in consideration of the objections, but PSAC did not back down, although the title of the report was changed from "The Hazards of Pesticides" to "Use of Pesticides." The battle between PSAC and Agriculture raged through the winter of 1963, but PSAC refused to compromise. The final report agreed with Carson's warning of the dangers of overuse of the chemicals and called for greater government regulation. President Kennedy ordered federal agencies to get involved, and Congress passed the first in a long series of environmental rules. Carson's book, despite its sometimes creaky science, has been credited with beginning the environmental movement in America, and partly that is because of the support Carson's premise got from PSAC. Historian Zuoyue Wang believed that report marked PSAC's height of influence in Washington.[20]

It also placed scientific expertise at the highest levels of defense research in the form of a director of defense research and engineering, Garwin wrote in an article for *Nature* in 2007. Herb York was the first; Harold Brown followed him. Garwin said that early on, PSAC was accused of being the scientific community's lobby inside the White House. Much of the opposition was channeled through Congress from scientists and from industry adversely affected by PSAC decisions. If your project was shot down, you went to see your congressman.

"Paradoxically, the higher profile of science and technology that the PSAC helped to create in government departments and agencies made it easier for its opponents to argue that the White House didn't need its own scientific advisers," Garwin wrote in the *Nature* article.[21] One day in the 1970s, a defense secretary, described by Garwin as "one of our most competent," told him he was not responsible if he submitted a program to the budget that was a clunker. The Office of Management and Budget would weed it out with PSAC's help. That was proof to Garwin that more expertise and influence in the White House was needed.

Since President Nixon killed PSAC, other presidents have brought scientists into the administration but not with the same power. In 1972, Congress created the Office of Technology Assessment, run by a bipartisan board with part-time consultants. OTA would invite passionate advocates and passionate opponents to serve on panels advisory to the studies, and have their opinions evaluated. In 1990, the elder George Bush created the President's Council of Advisors on Science and Technology. Then OTA was abolished by Congress in 1995, which some noted was like Congress shooting itself in the brain.[22]

"The United States has drawn particular strength from granting independent technical consultants access to government at many levels. Such individuals add knowledge, if not always wisdom. Today, that access is increasingly subject to political and ideological tests," Garwin wrote. He praised the Foresight program in the United Kingdom, with a sponsoring government minister for each project. "Foresight stands in sharp contrast to the absence of analysis in the United States for decision-making in and direction of programs such as missile defense, smallpox vaccine production and homeland security. The demise of integrity and competence in the US government is a matter for dismay."[23]

The current system makes use of the White House Office of Science and Technology Policy, and the director is John Holdren, who also is the assistant to the president for science and technology. He's a former president of the American Association for the Advancement of Science, has taught at Harvard and MIT. Holdren co-chairs (with Eric Lander) the President's Council of Advisors on Science and Technology (PCAST), seems to get high marks from scientists. He is an expert on population, energy, and the environment, and has been involved in disarmament, generally on the same page as Garwin. He served the White House during the Clinton administration as a key member of Clinton's PCAST. President Barack Obama appears interested in science, running a science fair for youngsters in the White House. Relations with the scientific community appear calm.

CHAPTER EIGHT
JASONS

Before Trinity, science had almost no influence on government or public policy. Thomas Jefferson sent Lewis and Clark west on a government expedition in 1804 to explore the continent, and they sent back tons of scientific samples to Jefferson's delight. Abraham Lincoln was actively involved in designing the *Monitor*, the first American iron warship in 1862. Theodore Roosevelt created the Food and Drug Administration, which hired scientists. But they were relatively small examples of government-science collaboration. After the explosion at Trinity, which clearly won the war and changed the world, now both government and industry wanted to hear what scientists had to say.[1]

Then the Soviet Union put a satellite, Sputnik, into space in Earth orbit. It didn't make things better when the Russians orbited a second, larger satellite with a dog aboard.

As part of the reaction, Americans saw science in a new light. Science in the Soviet Union now posed an existential threat. America needed to respond. Science reached an unprecedented level of appreciation and influence. A month and a half after Sputnik 2, President Eisenhower moved the Science Advisory Committee from the Office of Defense Mobilization into the White House—elevating it to the *President's* Science Advisory Committee. Isidor Rabi was the chair, and Hans Bethe was a member. They kept busy with things like antisubmarine warfare (which would become something of a specialty), chemicals in food—and disarmament. Shortly thereafter, the Pentagon set up the Advanced Research Projects Agency (ARPA) to investigate high-risk, out-of-the-box ideas for defense. Herb York was the first chair. That was followed in December 1958 with an umbrella organization, the Office of the Director of Defense Research and Engineering (ODDRE). The amount of money the federal government was now putting into science, thanks to Sputnik, was gigantic. Not only was the space race on, but the arms and science race was on as well. Physics, said Princeton theoretical physicist John Wheeler, "was booming."[2]

Meanwhile, PSAC was full of many of the same people who built the atomic bomb, and almost all came from America's elite research universities, especially from two institutions a few blocks from each other on the Charles River in Boston. It was decidedly an old boys' network—and it was largely made up of old boys, or at least middle-aged men. Wheeler, Oskar Morgenstern, and Eugene Wigner, all of Princeton (not on the Charles River), meanwhile formed an informal group they called the National Advanced Research Projects Laboratory, which later became the National Security Research Laboratory, to write proposals for government contracts. Murph Goldberger was a member.

The Charles River group began a series of summer studies funded by ARPA. The Pentagon would send them a problem, and then they'd meet during the summer and send a report back to the Pentagon. In the summer of 1958, they got twenty-two scientists together at the National War College in Washington. No one knew who decided who would get invitations, but it seemed to work. Defense officials were in charge of the briefings. Subjects included submarine detection (the Russians), submarine communications (the US), and the role of nuclear weapons in warfare. The Pentagon was very pleased. It was decided a national laboratory would be built, staffed with full-time scientists. But many scientists did not think that would work. What did work were the summer sessions by the Cambridge crowd. Another group, including Murray Gell-Mann, Goldberger, Keith Brueckner, and Ken Watson, tried to form a for-profit consulting group, Theoretical Physics, Inc., to the same end.

It was not that there were no scientists consulting with the government, but they were mostly working in agencies with political agendas. They were not independent. They were not inclined to tell their bosses what their bosses did not want to hear.

"We thought of starting a consultant group, a private business consulting group for the DOD [the Department of Defense]," said Gell-Mann. "We would work out mathematical problems of various kinds connected with equipment or various things. I was always interested in strategy more than anything else. But anyway we thought we would work for the DOD in this group and make some money and also do something to help the DOD and also to do something to help the country in the sense of pushing the DOD in useful directions instead of horrible directions."[3]

It never got off the ground. There was a need for independent science advice, so the compromise, which involved other scientists and especially driven by Charles Townes, was an informal group of invited scientists to act as advisors and problem solvers for the government. An advantage was that it would be able to bring in younger scientists as the Manhattan Project people were getting old. The organization would technically be under the umbrella of the Institute for Defense Analysis (IDA), a federal contract research center (FFCRC). The advantage over what already existed was that it would be entirely independent. IDA would simply be the money conduit.

It was to be called JASON. According to Goldberger, IDA had a colophon that was like a Greek temple, "and it was that that caught my wife's attention. Here was this group of brave young men setting out to conquer or defend the world. So [Mildred] just popped out the name JASON. That's the true origin of the name."[4] Despite almost always being written with uppercase letters, JASON is not an acronym for anything.

All the JASONs formed in 1960 and got security clearances, which at the time—in the height of the Cold War—wasn't a simple process. They would meet every summer, alternating coasts, break down into informal groups, and answer queries thrown at them by the government, mostly the Pentagon.

The first summer they discussed a subject that would dominate their research for years, antiballistic missile systems. One question put to the JASONs was whether or not a nuclear explosion in space would block the view of defensive missiles. If Russia, for instance, preceded a major attack by exploding a nuclear weapon in space, would it produce enough nitric oxide molecules to obscure the view of the missiles coming behind it? If there was enough of it, it would swamp the infrared from the rocket. The blackout wouldn't have to last long, just enough to make a defense worthless. The JASONs concluded that to do that would require a blast too large to be practical.

The question so interested Stanford's Sid Drell that he, at first reluctant to join the group, changed his mind after talking to a colleague, Mel Ruderman at Columbia. Not only was it interesting science, but it affected policy by encouraging the US Air Force to launch a satellite called MIDAS that would be immune to the blinding. The combination of interesting science and policy was, he told author Ann Finkbeiner, irresistible. It was there he met Richard Garwin, his friend and sidekick for the rest of their lives. The second summer tackled the issue of detecting decoys in incoming missiles.[5]

Membership was by invitation only, but that was an informal process. JASONs weren't sure how it worked. Initially, members were all physicists, but that gradually changed, and now there is a "sprinkling of chemists and mathematicians," Gell-Mann said. In the decades since, they have added modern biology, astrophysics, and computer science.[6]

"There was a real need for people who would do some work," Garwin told Finn Aaserud in an interview in 1986, "and furthermore it was clear that nearly all the people who were involved except for myself had been through World War II, either in the radar program or the atomic bomb program, so we needed to bring up new people with a familiarity with the defense programs and with clearances, so that they would be able to help the President's Science Advisory Committee. So in part it was conceived as a kind of training ground for panels of the President's Science Advisory Committee. In part it was designed to provide better studies which could be reviewed, and in part it was to get the Defense Department to do a better job itself of management in getting proper programs, because it would have these people with whom to interact."[7]

Greg Canavan, senior fellow and science advisor at the Los Alamos National Laboratory, who served as liaison between the JASONs and the Department of Defense in 1976, said he was told by Wheeler that as far as DOD was concerned, JASON was set up to help with the antiballistic missile issue. Part of his job was to defend them to DOD and to DARPA when they failed to come up with anything useful.

"I recall that they survived by a bit of a fluke," Canavan said. "Bill Perry, then undersecretary for research, happened to sit in on a Saturday morning [JASON] morning session. He liked the open discussions and ability to get concrete answers so he told DARPA Director [George H.] Heilmeir to keep them around."[8]

Garwin, who was too busy initially to become a part of JASON, was invited to join in 1967 and became its longest-serving active member. He had avoided getting involved with the earlier schemes at creating consulting companies. "I wasn't at all clear about the propriety of working with them," he said. By this time he had been on PSAC and consulting for the White House. He thought his doubts about consulting were not universally shared. A perceived conflict of interest doesn't stop a lot of scientists; JASON has a formal mechanism for preventing such conflicts. "I guess you can't complain about that. That's the capitalist system."[9]

"You can figure that any matter of great current concern to the Department of Defense will be the concern of the JASONs," said one government official.[10] The Defense Department was looking for someone to understand the antiballistic missile program, said Caravan, who was then with the Pentagon.[11]

A culture clash between the scientists and the Pentagon was inevitable and permanent. It is possible Garwin made things worse—or at least the young Garwin did.

"Originally there was a notion that anybody who was a member of industry would not be a member of JASON, that there would possibly be some conflict there," said Ed Frieman, a Princeton astrophysicist and JASON founder. "So Dick, at that point, was with IBM, even though he was at a very broad mandate from IBM. So he did not join JASON until some time later. . . . In those years—well, Dick of course was a member of PSAC and had been on just a huge number of panels. A few of us felt that . . . we should just wipe away the IBM proscription and just get him on board, that he was too valuable to have to worry about that kind of problem. And he was brought on board."[12]

In some ways, the creation of JASON was a generational shift as had happened at PSAC. Young scientists were replacing the Manhattan Project men.

"A number of us expressed concern that the only people we ever saw when we went to these committees were A, the same people, and B, they were old-time Manhattan Project warriors and it was time for a new generation to step up and take some responsibility," Goldberger told Dan Ford. "In a complex way . . . there was born the notion of the JASON group. I, together with a man named Kenneth Watson and another named Keith Brueckner, were the ones who really put this together, though the fundamental impetus came from the person who was director of research at IDA, the Institute for Defense Analyses, Charlie Townes. . . . We chose the people who formed the initial JASON group. We did not choose Dick because he was working for a company, and we didn't want there to appear any kind of conflict of interest or what have you. We wouldn't take anyone who was at the national laboratories at the time, either."[13]

Goldberger eventually changed his mind and recommended Garwin because of his work in theoretical physics. JASON is largely made up of theoretical physicists, because the founders envisioned that most of the problems they would be asked to solve would be best approached theoretically, although there might be experiments and practical applications.

"What they needed was a set of interpreters, people who could really provide some insight, understanding and go part of the way toward a theoretical posing of problems," Garwin said.[14]

Garwin had briefed JASON before he became a member, mostly about antiballistic missile problems, something he had studied with

PSAC. One way to hide missiles from an enemy, he told them, would be to launch lots of balloons and hide a missile in one of them.[15]

Once in JASON, Garwin was almost immediately nominated for the Steering Committee that runs the organization. The chairman of that committee is the nominal chairman of JASON.

Goldberger said Garwin might not have been entirely well-suited for the job. Garwin had little patience with letting people labor on problems where he could see the solution instantly. He was elected for a three-year term but only lasted two.

"The troops were restless," he said. "I was a fount of innovative management." One of his ideas was to encourage 10 percent of the JASONs to resign each year to bring new blood into the organization. That was not popular. Nor was an edict for an annual self-assessment. "I think only one or two people actually complied."[16]

The first paper with Garwin's name on it was in 1967, "JASON Review of STRAT-X Report," having to do with the MX missile system, land-based and sea-based missile systems. The US Navy was considering—and Garwin supported—a weird system for carrying missiles. The prevailing notion was that in order to carry large missiles, the navy would have to build huge submarines. For the missiles to be fired they would have to stand upright, which would require a pressure hull forty feet in diameter, as wide as the missiles were long. What if the missiles were in canisters lying outside the sub that would be swiveled upright when they needed to be fired? The size of the submarine would not matter. This system, the Undersea Long-Range Missile System (ULMS), was never built. But in 1980, Garwin and Drell led a JASON study pushing the ULMS concept in a new direction—to have the land-based ten-warhead MX missile carried in two or four capsules strapped to a tiny submarine. But the navy would have none of it and criticized the report on grounds that the authors did not know enough about submarines. Sid Drell, who would become the Sundance Kid to Garwin's Butch Cassidy, eventually went to William Perry, director of research in the Carter administration, to get a letter assuring everyone that Drell and Garwin really knew enough about submarines to be listened to. Garwin would continue his expertise on submarines, testifying to the British Parliament in 2007 on determining the life expectancy of their Vanguard class nuclear-armed submarines. Garwin and his colleagues concluded they lasted longer than they were expected to when they were built, and good to go into the 2020s. Drell has a model of the little submarine in his Stanford office.

Garwin is also the originator of a plan called Fish RAGU. The plan was for a fleet of little battery-powered "fish" to swim along with a sub.

"At low speed it will swim around all day and could be used for all kinds of things, like going up to the surface and listening to the radio there and then sending the information down on megahertz acoustics links which cannot be heard at any significant distance because of attenuation," Garwin said. "You can go out and you can check the submarine radiated noise and do useful things as sonar auxiliaries." When it was obvious the MX system required reliable communications with the submarines, Garwin and the JASONs figured the "fish" would be a perfect vehicle. They would submerge a few feet below the surface and use a ferrite-rod antenna to receive a signal from command on shore, either AM or FM. Or, the sub could deploy buoys connected to the sub with fiber optic cable to handle the transmissions. The fish would be housed in the sub, would weigh 50 kg. (110 pounds), and would recharge and could travel for about an hour at ten knots.[17]

Then there was the paper on "bombs that squeak." What if the US had to attack Russia with nuclear weapons? How would we know which targets were destroyed and which weren't?

"Well, all you really need to do is to observe our own nuclear weapon explosions. You know exactly when and where they are supposed to happen and so you only need to look to see whether it did happen and to confirm the location to hundreds of meters," Garwin said. The plan was to equip every nuclear weapon with a high-explosive powered generator of microwaves that would give a coded pulse to a special set of satellites that could measure exactly where the bomb went off.[18]

How much of JASON's work is classified is not known, but it is probably more than half. At JASON meetings, announcements are made before sessions on what level of classification would be required to sit in. A guard stands outside each meeting room. JASONs and visitors wear name tags with the level of clearance displayed. The guards know who is allowed in and who is not. About half the sessions are not classified, but attendance is limited to government employees or contractors. The spring 2016 JASON meeting in Falls Church, Virginia, included sessions on gravitational waves, enabling data-driven government, and black hole physics, which did not require a clearance. On the other hand, the schedule also included one on Russian nuclear issues, the Lebanese Hezbollah's evolving capabilities, and Russian strategic offensive strike systems, all of which required a high clearance level to attend. The main sessions, during the summer at La Jolla, California, run somewhat like that, but instead of two days they last six weeks or more. There are no known incidents of classified material leaking from a JASON meeting or a JASON.

Garwin said one of the reasons JASON is successful is that the US classified materials are more widely shared than such materials in Britain or France. "There is a matter of critical mass," he said. "You have to have a certain size in order to make an impact across the spectrum."[19]

Garwin is a bit cynical about how it works, having spent so much time dealing with the government. "Everybody [government agency] gives money and they understand that the way we work, we have to spend money on meetings, inspection trips, general briefings, and so on. It's only a question of whether in the end they are repaid commensurate with the expenditure. Usually we will negotiate, resulting in an agenda for the year, topics to work on for each of the sponsors. So if we work on any of those and deliver, that's OK. Sometimes we will be in a position where we agree to do something, some particular thing, and then if we don't deliver insights and reports on that it's not good for us or for the people in the agency who have stood up for JASON. Finally, it's easier for an agency head or somebody high up in the agency to tell people 'bring problems so that we can give them to the JASONs because we are paying them this money anyhow.' That's easier than telling the people down in the ranks to 'bring problems and money, because there are these smart people who might help you solve your problems.'[20]

"Many people don't want their problems solved. Their problems are their security, their paycheck. If they did want their problems solved they have people with whom they are more familiar that they're likely to go out to and try to get to help solve their problems. It's not always an easy thing to have meaningful problems posed with enough freedom so that we can look at them.

"Agencies will say they want independent judgment, but many do not. They want support for their projects and theories. Sometimes clients will not give all the facts, using classification as an excuse. The Navy, which seems to have tighter restrictions than the other Armed Forces, is particularly bad that way," Garwin said. And as the government has grown, the situation has gotten worse. Because of the way JASON is set up, there is compartmentalization. For many years a group of JASONs worked on both detecting and hiding submarines, and the group doing that research was distant from other JASONs.[21]

Garwin's public stances on subjects, including those before JASON, were not appreciated by everyone. "Yes. It's been a problem," said oceanographer Walter Munk. "We have some people who have a very strong public presence. Our most famous member is Richard Garwin. It's counter to our, I think, general taste—my taste, not everyone's—that we ought to do our job and keep our mouth closed. I think Dick represents a very responsible member of US society, and if he wants to speak, he should. He has been careful, of course, not to speak on matters that he's not allowed to, or he'd be in jail. . . . Well, I think people have done the right thing. They said, 'We wish he'd keep his mouth closed, but as long as he stays within certain accepted rules, we will defend his right to do so. And we've stuck to that.'"[22]

CHAPTER NINE

VIETNAM AND McNAMARA'S WALL

Vietnam was a nerd-driven war. Robert McNamara, who was a nerd, was certainly on their side. The military's relationship with the nerds was mixed. To some extent the military was onboard and did what the nerds advised. Or the military did what it was told but didn't like it and was passive-aggressive about it. Or there was real conflict. At the core of much of it was operations research, applying real data to modifying or instigating action. And of course a military reluctant to do what civilians say they should do.

Wading into that morass were Garwin and the JASONs. Some of what they advised was reluctantly accepted; some was simply rejected, and some was passively ignored.

Nonetheless, it is certain that Garwin and the JASONs helped advance the concept of the electronic battlefield in Vietnam. In the process, it brought them notoriety, conflict, and led to the kind of public exposure the organization had avoided since its founding. In some ways, it was the high point in JASON's influence with the US government, and it was born out of failure, largely because the armed forces were dragged reluctantly to a plan not of their own inventing and resisted implementing it whenever they could. Defense Secretary Robert McNamara hoped JASON's ideas might save America's soul. For Garwin, who was in the middle of it, the JASONs were just doing what they were paid to do.

The Washington-science establishment had been doing some work for the Pentagon on Vietnam, especially PSAC, and that included Garwin. But the Washington establishment was not capable of coping with grand plans. They were, Garwin said, more incremental. The JASONs thought big.

Garwin said that PSAC operated in a different milieu than the Washington community. The scientists complained how bad the intelligence was from Vietnam. "In Washington you are really isolated. People who read newspapers often get a better understanding of these things than people who sit in Washington to process intelligence," Garwin said.[1]

The JASONs were not so much asked to help but injected themselves into the stew. They thought the military needed them. That was unusual for the group, who rarely do things on its own. The JASONs generally do what they are asked to do, and the military didn't think it needed them.

At one time Garwin, as a member of PSAC, wrote Chester Cooper, a "very capable person"[2] on the staff of McGeorge Bundy, the national security advisor, volunteering that the JASONs had considerable experience and information that might be of use in the war.

"We had all this experience, which he didn't know about. He had been for months across West Executive Avenue in the National Security council staff and he had no idea that there were people who had good channels for finding out, who had a lot of information, and who had made recommendations. It's just too bad that the government doesn't make better use of the information that is available," Garwin said.[3]

"JASONs is a group of technical advisors and the whole point of JASON is that we have people of all kinds of political opinion who work together," said physicist Freeman Dyson, who came close to resigning over the ensuing political debate. "The whole purpose was to not get involved in philosophical arguments. It's details. It's what we do. JASON is an activity. They come to us with problems. We come to them with problems. We have to agree. Whether they actually do anything or not is not our responsibility. We do what we can do."[4]

"What we should have told McNamara at the time was to take a flying jump," Murph Goldberger told author Ann Finkbeiner. The intrusion of JASON in the war was, he said, "almost a textbook demonstration of the arrogance of physicists."[5]

The JASONs eventually would be getting $1 million a year for their work in Vietnam. "The United States' decisions, in the early months of 1965, to launch a program of reprisal air strikes against North Vietnam's evolving progressively into a sustained bombing campaign of rising intensity, were made against a background of anguished concern over the threat of imminent collapse of the government of South Vietnam and its military efforts against the Viet Cong," the authors of the Pentagon Papers stated in a convoluted sentence only a bureaucrat could write. The bombing operation was called Rolling Thunder. The bombers were sent essentially to destroy an agricultural society using inappropriate modern weapons.[6]

The US military decided that the only way to win the war was to cut off the supply lines from the North, the so-called Ho Chi Minh Trail. Rolling Thunder attacked the trail with gigantic bombardment. The first mission was March 2, 1965. Despite the ceaseless fall of explosives, the North showed no sign of submitting.

A month after the bombing started, Garwin began discussions on the tactical use of aircraft in Vietnam at JASON meetings and to plan for using Vietnam as something of a laboratory for improving the planes. What better way to study the design of combat aircraft than in combat? About a quarter of PSAC's discussions, with him as an instigator, concentrated on the war.

History—and perhaps arrogance—impelled the JASONs. Science had in the end won World War II. *We are scientists. We are the cream of America's physical scientists.* Some of the JASONs said it was time that they step in and solve the problems for the military. They didn't seem to realize that the military was not interested in having them solve their problems.

In 1964, Murray Gell-Mann called a JASON meeting and invited experts to come and lecture on what was happening in Southeast Asia. The idea that they should intervene with expert advice probably was reinforced by what they were hearing. Many of the JASONs were unimpressed with the experts, and although they issued two reports from the sessions, nothing came of them.

One fact confronted them. In one of the most famous sentences to come out of the Pentagon Papers, the futility of the bombing campaign was clear. "As of July 1966, the US bombing of North Vietnam has had no measurable direct effect on Hanoi's ability to mount and support military operations in the South at the current level."[7] McNamara totally agreed. The military did not.

Edmund Burke once wrote that those who don't know history are doomed to repeat it. Those who do know history, it may be added, are doomed to stand by while those who don't repeat it. The whole rationale behind Rolling Thunder was the premise that if you bombed the hell out of the North, the people would finally have to submit. The idea that you could bomb someone into submission had, by that time, been totally disproven to everyone outside of the Pentagon who had a passing familiarity with history. The Germans tried it with the British in World War II, and all it did was bond the British people and produce memorable photographs of Winston Churchill standing in the rubble of London. Both Germany and Japan had been turned into piles of debris. Even after years of bombing, including firebombing wooden Tokyo, it took the complete destruction of two cities by atomic bombs to get the emperor of Japan to order a surrender, and that was over the objections of a military that still wanted to fight on. Germany only quit when Russian troops poured into Berlin at a time when its remaining cities had ceased to exist as cities, the government no longer existed, and much of the population was homeless and hungry. Hitler was arming children to defend the capital because they were all he had left. North Vietnam, not an industrialized society with large targets that could be destroyed, was even less likely to collapse under bombardment. Mostly what you did with intense bombing was kill people and make the rest angry. Additionally, the Vietnamese had a long record of successfully fighting colonial oppressors. But as the supplies and men moved south, the US Air Force increased the number of bombing missions. McNamara, convinced that Rolling

Thunder was worse than useless, wanted a substitute he could force on his military, one more effective and morally acceptable. So, when the JASONs got involved, he was an enthusiastic listener.

The JASONs recommended that their proposal be implemented by a unit made up of all three military services, which was code-named the Defense Communication Planning Group (DCPG), which in turn had an advisory committee made up in part of JASONs, including Garwin. George Kistiakowsky was the chairman.

McNamara had been Kennedy's defense secretary and, like most members of that administration, stayed on with the Johnson administration after Kennedy's assassination. JASON's best idea at the time was to expand the use of gunships, cargo planes, with cannons aimed out the side doors against the trail. Targets on the ground were stationary, and it was difficult to hit a high-flying plane a mile away over the jungle, giving them a distinct advantage. JASON recommended having nineteen gunships diverted to interdict the supply line from North Vietnam. David Packard, the deputy secretary of defense under Nixon, thought that was a grand idea, but there was no way, he said, to get the air force to divert nineteen planes for this tactical effort. There were not many of the planes, they were being used elsewhere, and were a scarce resource. He probably understood that the air force was unlikely to embrace an idea that came from outside its branch, particularly from civilians. He offered to expedite the production of new C-130s, but of course that would take a while.

McNamara, former head of Ford Motor Company, was getting sick of the war and unilaterally declared—for budgeting planning purposes, he said—that the war would end in 1965. The fighting, of course, wouldn't really stop, nor did the money, but he wanted people to get used to the idea of winding down support efforts and to get them used to the idea that the war would end soon—he hoped. That decision essentially ended military research and development for the war. But it left the door open for non-Pentagon agencies.

In 1965, Garwin's PSAC Military Aircraft Panel invited briefings from, in sequence, the chiefs of staff of the army, navy, air force, and the marines. The officers told the committee their wish list. The marine general said he wanted rockets so he could quickly transport his troops wherever they needed to be, apparently faster than the current aircraft. "And we asked him please to get real. What could be done before the war might end," Garwin said.[8] The general said he would like to know where his men were so that they would not be accidental targets for American canon fire—friendly fire. Garwin told him they could do that in a month. And they told him how, but it didn't happen.

The military had bulky LORAN receivers, a form of radio, in vehicles and at bases all over Vietnam. LORAN was a radio navigations system first used in World War II. The squads all carried small high-frequency radios to communicate with each other. By fitting the radios with a small plug-in attachment, the radios could pick up but not interpret LORAN signals and send them to the fire base. In turn, the fire base would feed the received signal into their bulky LORAN receiver to determine the squad's location so that the artillery would be aimed elsewhere. The conversions of each radio would take minutes and the overall development a few weeks.

The panel's advice was not taken. Nothing was converted. On the other hand, Garwin told David Kestenbaum at NPR that at a day-long meeting convened by PSAC to evaluate the idea and alternatives, the navy had six laboratories and had eight proposals for monitoring troop location. "By the time the Vietnam war ended seven or eight years later, not one of them had been implemented. If they had just taken our approach, which was eminently doable, maybe not optimum, they could have [a system] within two or three months."[9]

(In 2008, the US military began deploying Blue Force Tracking, a system based on GPS satellites—using some of the technology invented by Garwin—to keep track of troops in the field. It not only monitors friendly forces, but can keep track of enemy troops as well. Its record in the Iraq War is mixed, and the question remains if the JASON idea wouldn't have been better, faster, and cheaper.)

Relationships between the nerds and the military were sometimes rocky, and Garwin's personality didn't make things easier. "A problem with Richard, I noticed in policy questions having to do with military matters and so on, a problem with him was always that if the system somehow included people then his comments on the system were not always so wise," said Gell-Mann. "If it was a purely technological question then he always had a brilliant solution but to the extent that there were human beings in the loop his recommendations were sometimes not very sensible at all."[10]

"I don't like bureaucracy, I am not particularly good at it," Garwin said. "I worry more than people should who are managers and I have many other things to do. If I am not essential in such a role I try not to do it."[11] Garwin said that to humans there are five essential elements to life, and they are "food, water, shelter, sex and not being told what to do."

Garwin was full of ideas, Gell-Mann remembered. "He helped a lot. He made some very intelligent suggestions."[12] He recommended that some of the bombing make use of drones, then in their early stages of development. That went nowhere because it did not involve "heroic pilots." People joined the air force to fly, not to sit at a desk and command a drone; they joined to fly aircraft large or fast, engage in dog fights with fervent enemies, scarfs metaphorically blowing in the wind, rescuing beleaguered infantry. Think Snoopy in his imagined battles against the Red Baron.

He missed the human element.

Garwin also envisioned a drone helicopter and the development of what are now called smart weapons.[13]

McNamara had been sent a four-volume report on Rolling Thunder, written by a group that included many JASONs. The report was largely the product of Gordon J. F. MacDonald, a geophysicist and head of the JASON committee on Vietnam, along with IDA. MacDonald concluded that the bombing was ineffective for cutting supplies above a particular latitude line. If the supply lines were to be interdicted, it had to be farther south. Bombing Hanoi was useless, he said. That supported McNamara's skepticism about the bombing. McNamara wanted some ideas and proposals to replace Rolling

Thunder, and urged President Johnson to halt the bombing. Johnson would have none of it. "I basically do not regard bombing as a matter of science," Johnson said.[14]

MacDonald brought the Vietnam issue to JASON.

In 1966, with Gell-Mann pushing, Murph Goldberger, then chairman of JASONs, decided the most useful thing the group could do would be to find a way to block torrents of men and supplies heading from the North to the South along the Ho Chi Minh Trail. He put the topic on the agenda for the summer JASON meeting at the University of California, Santa Barbara.

Meanwhile, back on the East Coast, the research clique at Cambridge, the Charles River gang, led by the Manhattan Project veteran Kistiakowsky and including Wiesner and Zacharias, decided they wanted to get involved as well. They had the same motivation. They were physicists, and they were going to clean up the trail problem. With the Defense Department's approval, the Ho Chi Minh Trail went on the schedule for their summer meeting. The two groups worked separately, JASONs East working out of a girl's school in Wellesley, Massachusetts, and JASONs West at UCSB. Garwin, who had just joined JASON, was JASONs West. He had done some work on Vietnam as a member of PSAC and the Military Aircraft Panel. He wanted to bring the JASONs up to speed on what was happening in Southeast Asia. Even Washington was behind in understanding the issues, he felt. There were JASONs in both groups. IDA provided the supporting bureaucracy.[15]

A conclusion reached by the JASONs East, transmitted to McNamara, was that the JASONs were singularly unimpressed with the intelligence the Defense Department was getting from the military in Vietnam, and in general they seemed clueless about everything from enemy casualties (in many cases they were exaggerating totals) to about whether all the bombing would have a beneficial effect on the outcome of the war. The JASONs issued several reports, one calling for an extensive string of sensors that would enable the air force to know where to send bombers to cut off the trail, and another that supported the idea that Rolling Thunder was a failure.[16]

Another report from JASONs East was for a barrier across the trail made of barbed wire, obstructions, defoliation, and fortresses. The Ho Chi Minh Trail itself was bifurcated, with different road-ways for men and for vehicles. Because of geography and politics, the fastest way south from North Vietnam passed through a portion of Laos before slanting back into South Vietnam, the result of how the borders had been drawn by the French. By treaty, the US could not put ground troops in Laos, although Garwin pointed out that the CIA had people there. That meant the Laotian portion of the trail could only be attacked from the air and, for diplomatic purposes, would remain secret.

The two JASON groups were not coordinated. For one thing, JASONs West was not formally invited to participate with JASONs East. Garwin kept himself informed what both groups were doing to make sure they didn't make any mistakes or to see if he could find an opportunity to do something.[17]

The two groups came up with two very different plans. JASONs East would have a conventional physical barrier, with razor wire, mine fields, fortresses, artillery for cover, and every physical barrier they could think of. That might work on the Vietnamese portion of the trail—across the "demilitarized zone" (DMZ). The West came up with an air-supported electronic barrier for west of the Laos-Vietnam border.

There was precedent for such a structure, the Morice Line built by the French in the 1950s during the Algerian-French War. It con-sisted of a physical fence that ran 285 miles along the Tunisian-Algeria border and 435 miles along the Morocco-Algeria border to keep Algerian nationalists from crossing into Algeria. It had mine-fields on either side, and the French added sensors on the fence, including some linked to artillery units. It worked.

McNamara passed both JASON plans up through the chain of command, and they were immediately shot down by the Commander in Chief of the Pacific (CINCPAC). It would tie down too many troops and supplies, all of which was needed elsewhere.

Garwin found the JASONs knew nothing about land warfare,

having spent their time studying missiles. He found thirty ways of getting around the barrier the JASONs had been planning—ways the North Vietnamese could beat the fence and foil the technology.[18]

The JASONs West plan was called the Air-Supported Anti-Infiltration Barrier. No forts, no fence, no obstacles. The air force would drop noisemakers to the ground. When trucks or people triggered the noisemakers, detectors in the trees would signal planes in the air that would bomb the area the noise was coming from. It was devilishly simple, sending toys to war. Would it work?

Garwin had worked on antisubmarine warfare techniques using sonobuoys, devices dropped into the water to listen for the sound of enemy subs. The sonobuoys would deploy, with one part, equipped with an antenna, floating at the surface, and a second part, with a hydrophone, sinking hundreds of feet deep, listening for submarines. Garwin suggested that these sonobuoys be the basis of the JASON sensors. What they heard was relayed to P-3 aircraft circling above. The plan was to modify this process for the jungle. The sonobuoy derivatives to be used in Laos would be air dropped and slowed by parachutes, listening to noisemakers strewn on the ground to relay what they heard. Some noisemakers were small firecracker-like devices. Others were bags of explosive plastic beads that probably wouldn't kill anyone but could injure and maim.

The detectors were designed to capture noise from either men or trucks and relay the data. Aircraft circling above at 20,000 or so feet—piloted or a drone—would get signals from the detectors and send data via microwave to a computer on the ground in Thailand. After filtering out the sounds of the forest—birds and animals—it would pinpoint targets and send the bombers. The technology was sensitive enough to pick up and relay the conversations of the Vietnamese as they walked by. The sonobuoys were also designed to pick up magnetic fields, vibrations, and even certain odors such as urine.[19] The computer in Thailand would give the coordinates of targets, and circling aircraft would attack the coordinates without ever seeing what they were bombing through the jungle canopy.

The Defense Communication Planning Group (DCPG), one of the code names for the air-supported barrier, even went into the jungles of Panama to see if the sensors would really work among all those trees. They did.[20]

"Sensors don't keep anybody from coming through," Garwin said, "so the idea was that you would have such an effective capability of striking trucks or whatever in response to sensor indications that they wouldn't come at all. It's like a perfect minefield or a fence; there's no sense coming, you won't get through; so you don't hurt anybody."[21] In the end, it would be more humane.

The air force, as usual, resisted. "In order to be most effective," Garwin said, "one needed to have the strike aircraft under the same command as the sensors. The Air Force isn't organized like that."[22]

The air force had research laboratories, and researchers of their own, and this plan, ridiculously simple on the face of it, did not come from their shop. One excuse: air force high-speed aircraft were incapable of dropping the sensors with sufficient accuracy. Eventually, the navy stepped in with planes from carriers, further infuriating the air force. Originally, the air force wanted to fly a Lockheed Constellation over the trail, a graceful aircraft that was a popular commercial, propeller-driven airliner. Eventually and reluctantly, they sent up small private plane–type aircraft. When the plan was finally put in place, forced on them by McNamara, it worked, except that instead of sending planes immediately, to bomb the targets, the air force waited a day to merge the data from the sensors with other intelligence. The planes showed up a day late, when the trucks or men were long gone. When they did send out aircraft, they dropped cluster bombs, which were less effective in the geography. They were bombing the highway instead of the convoys on the highway, Garwin said, and with the wrong munitions.[23] It didn't help that the head of the DCPG was General Alfred D. Starbird, head of the Pentagon's defense communications agency—an army officer rather than an air force one.

The fence would have to be constantly replenished and revised. Garwin's list of ways the North Vietnamese could evade the sensors sent the JASONs to work on counter-counter measures. The total estimated cost of the line was $800 million a year. The JASONs had no illusions this would end the war; its purpose was to simply slow down infiltration and disrupt the rhythm of the conflict. The motive, JASON Hal Lewis told Finkbeiner, "was as pure as the driven snow."[24]

The report was delivered to McNamara on August 15, 1966. He had read the briefing documents and had questions. The JASONs understood from the meeting that McNamara's most important goal was to stop the bombing. On September 3, McNamara took an army helicopter to Zacharias's summer home on Cape Cod for a final briefing. One of the questions he asked was whether the North Vietnamese were likely to take some of the steps Garwin had suggested to get through the barrier undetected, but was told by the military they were more likely to keep going and take the casualties.

The North Vietnamese did, as Garwin predicted, find ways to evade or distract the monitors. Somehow, they discovered the smell sensors, which were designed to detect urine, so they sent out women with buckets full of urine to places away from the trail. The sensors picked up the smell, the relays sent the data to the circling planes, the planes passed the information on to the computer in Thailand that sent the jets, and the US Air Force bombed buckets of piss.[25]

Animals also threw the network into confusion. Every time an ox tripped or passed gas, the alarms went off.

The military remained hostile, with CINCPAC again totally opposed. McNamara had ordered them to build it anyhow. The JASONs were a bit surprised. They thought—as scientists always do—more research was needed before the thing could be built and that McNamara was rushing things. Part of the opposition from the military was the ghost of the Maginot Line that the French built to defend against a German attack before World War II. The Germans simply went around it. The military feared that the North Vietnamese would do the same—through Thailand.

All US efforts—which at one time included artificial rainmaking to produce mud—were to no avail. Men and equipment kept flowing.[26]

The barrier became known as McNamara's Wall or Fence. The cost now was up to $1 billion.[27] McNamara eventually would resign over the issue of Rolling Thunder and the way the war was being fought, and credited the JASONs with providing some of the information behind his decision. The Pentagon would not let it go at that. One of the official histories of the war, Edward Drea's *McNamara, Clifford, and the Burdens of Vietnam*, issued by the historical office of the secretary of defense, summed it up:

[The wall was] a metaphor for the secretary's arbitrary, highly personal, and aggressive management style that bypassed normal procedures and sometimes ignored experts to get things done. He had adopted an idea from civilian academics, forced a reluctant military to implement it, opted for technology over experience, launched the project quickly and with minimum coordination, rejected informed criticism, insisted available forces sufficed for the effort, and poured millions of dollars into a system that proceeded by fits and starts.[28]

Gell-Mann, among others, regretted the work the JASONs did on Vietnam. "I was stupid. It was really stupid," he said. "They promised us that it was going to be a replacement for the aerial bombardment of North Vietnam but instead it was an add-on which was a direct violation of their contract with us and their promise to us."[29]

For many of these men, this was doubly painful. "Twenty years after their handiwork wrought destruction in Hiroshima and Nagasaki, Manhattan Project veterans found themselves ensconced in key government and military advisory positions just as the war in Vietnam reached lethal new heights," the historian Sarah Bridger wrote. "Perhaps no other group of Americans offers as poignant a portrait of the tortured path from Hiroshima to Hanoi as these men do. Twice in their lifetimes, these elite scientists willingly contributed their expertise to devastating projects whose outcomes they couldn't control and would come to regret."[30]

Sensors similar to those invented for Vietnam were put to service in the Gulf War thirty years later.

If the wall itself was a flop, what happened in a region of Vietnam during the winter, spring, and part of the summer—five months and eighteen days of 1968—was historic, and may have changed warfare forever. Thousands of US Marines were surrounded and nearly overrun by the North Vietnamese army but were saved by JASON's technology.

Khe Sanh was in an area close to the Vietnam-Laotian border, inhabited mostly by an ethic group called the Montagnards. US Special Forces had built a base then near a French fortress, part of the plan to interdict the Ho Chi Minh Trail. An airstrip was built nearby after it was decided an invasion of Laos would be necessary. In October 1967, the North Vietnamese began a ground offensive against the marines stationed in the area, firing as many as 100 to 150 rockets at the base a day, along with mortars and artillery shells. They closed the only road out, Route 9.[31] The marine high command wanted to abandon the area, but General William Westmoreland insisted it was vital to remain. Part of the inducement was the mass of North Vietnamese troops that had moved into the area. No one knew why the North Vietnamese wanted a battle there, but Westmoreland could not resist. He wanted a conventional battle, almost a rarity in Vietnam, as he thought he had the firepower to win.

In September, the air force began a bombardment of the North Vietnamese positions that eventually grew into one of the most intense aerial attacks ever launched on such a small area, totaling the equivalent of five Hiroshima atomic bombs, the aptly named Operation Niagara. Even naval ships joined in the battle, its big guns firing from the sea. On the ground, some of the battles were among the fiercest of the Vietnam War. One entire marine brigade was effectively destroyed.

When it became clear that the marines were surrounded, Westmoreland and some other top officers considered using low-yield nuclear weapons to get them out. For instance, a small tactical nuke dropped on one of the mountain passes leading to the Ho Chi Minh Trail would block it. Scientists in an advisory capacity knew of the request and were appalled. The JASONs actually produced a report looking at the potential, but described the idea as at best useless even if they dropped ten nukes a day. Kistiakowsky, who opposed the war in general, sent a telegram to former president Eisenhower, asking him to tell Lyndon Johnson, who was visiting Eisenhower in California at the time, to reject the notion. McNamara killed the idea with the excuse that the geography of the hilly area would render the weapons ineffective.

On January 31, the North Vietnamese began the Tet Offensive, and massive numbers of North Vietnamese troops moved into Khe Sanh, reinforcing the siege. The 5,000 marines were now surrounded by 20,000 North Vietnamese. The only way out was by helicopter, and the helicopters were under constant fire. It looked all the while a repeat of the historic French defeat at Dien Bien Phu, which effectively ended France's Southeast Asia colonial empire. There were differences: Dien Bien Phu was in a deep valley, and the Viet Minh took the high country, leaving the French at the bottom, while Khe Sanh was on a hilltop. Helicopters, a relatively new invention, were not available to the French, and men and supplies were parachuted in. The French missed the target, and the soldiers and material were captured. On the other hand, Khe Sanh was within artillery range; Dien Bien Phu was in the middle of nowhere.

The same general who beat the French at Dien Bien Phu was now running the operation at Khe Sanh and using the same techniques, digging trenches and tunnels and moving incrementally toward the Americans.[32]

Garwin and MacDonald flew out to see the situation in February 1968, but because of the Tet Offensive they could not land in Saigon, so the plane went to Bangkok instead. MacDonald, however, found a helicopter and got into besieged Khe Sanh, "a real act of courage," Garwin said.[33] In Thailand, Garwin could put on a headset and listen to the North Vietnamese soldiers talking about how they had lost four trucks. He watched a television screen with the sensors marked at their locations. If the sensors detected anything, they changed color. It was possible to watch the trucks go down the road, triggering sensors in sequence.

Back at home, however, all hell had broken loose. An anonymous caller told a staffer of the Senate Foreign Relations Committee that the committee might inquire why experts in nuclear weapons were flying to Vietnam. The story got leaked to newspapers, and Washington was abuzz with the rumor that the Johnson administration was considering using nuclear weapons. The Pentagon denied that was the case. Garwin couldn't respond because the barrier was then a secret. He admitted later he had probably told someone of his forthcoming trip when he should not have.[34]

The marines, still under constant fire, could hear the tunneling. The high command ordered the sensors shipped into the besieged post. They were still technically a secret, although by this time not to the North Vietnamese who were essentially tripping over them.

On the night of February 3, according to a letter written by a marine captain, Mirza Munir, the sensors on a ridge behind one of the hills triggered an alarm. The enemy was moving toward the hill. The next night the sensors went off in other places, and Munir surmised the enemy was moving. Working from maps and plots of the sensors, he figured out exactly where the marines should aim their artillery. The computer in Thailand recorded the voices of panicked Vietnamese fleeing the battlefield as the shells made direct hits on their positions. A month later, the North Vietnamese tried a major attack on the marines only to run into a jungle of sensors and death. The use of the sensors eventually improved to the point that they no longer had to rely on the computer in Thailand; the marines had portable monitors. The siege was broken, and the marines were able to truck out on Route 9. And as is not unusual in war, Khe Sanh was abandoned after all that effort and lives.[35]

The JASON devices are widely credited with saving the marines, but it is impossible to know for sure, nor is it possible to estimate, how many marines' lives were saved, if any. Marine officers, testifying before Congress, said they thought the sensors turned the tide and

Marcel Weinrich, Leon Lederman, and Richard Garwin (left to right), with the equipment used to demonstrate parity violation in regard to the muon. January 1957. *Photo courtesy of the Richard L. Garwin collection.*

President Nixon meets with the President's Science Advisory Committee (PSAC). On his right, his science advisor, Edward E. David. On his left, PSAC vice chair John Baldeschwieler. Garwin is at the far end of the table. February 23, 1971. *Photo courtesy of the US government.*

PSAC meets with President Lyndon Johnson. Garwin is at the front, with his hand on a chair. March 21, 1966. *Photo courtesy of the US government.*

Garwin in his IBM lab, with a seventeen-inch touch screen. March 1986. *Photo courtesy of IBM.*

IN HONOR OF THOSE MEMBERS
OF THE CENTRAL INTELLIGENCE AGENCY
WHO GAVE THEIR LIVES IN THE SERVICE OF THEIR COUNTRY

Richard and Lois Garwin at CIA HQ for receiving the R. V. Jones Award for Scientific Intelligence. March 1996. *Photo courtesy of the US government.*

Dinner party at the home of Evelyn and Bob Frank. Lois and Garwin (back row, to the right from lamp), Norris Bradbury (middle row, first on left on sofa), Bob Frank (middle row, on sofa, third from left), Richard Feynman (on Frank's left), Evelyn Frank (on Feynman's left), Hans Bethe (front row, right, sitting on chair), Luis Alvarez (second from behind Bethe, sitting back), Rose Bethe (middle row, far right, kneeling behind sofa), Rudolf Peierls (back row, third from right), Harold Agnew (back row, sixth from right), Beverly Agnew (on Harold's right), and others. April 1983. *Photo courtesy of the Richard L. Garwin collection.*

WESTERN UNION
TELEGRAM

W. P. MARSHALL, President

CLASS OF SERVICE

This is a fast message unless its deferred character is indicated by the proper symbol.

SYMBOLS
DL=Day Letter
NL=Night Letter
LT=International Letter Telegram

1201 (4-60)

The filing time shown in the date line on domestic telegrams is LOCAL TIME at point of origin. Time of receipt is LOCAL TIME at point of destination

1962 JUL 25 PM 1 58

SYB060 SSD204

SY WA244 GOVT PD RX= THE WHITE HOUSE WASHINGTON DC 25

:RICHARD L GARWIN, REPORT DELIVERY, DONT PHONE=104P EDT=

16 RIDGECREST EAST SCARSDALE NY=

I HAVE SIGNED YOUR COMMISSION APPOINTING YOU A MEMBER

TO THE PRESIDENT'S SCIENCE ADVISORY COMMITTEE.=

IT GIVES ME A GREAT DEAL OF PLEASURE TO DO THIS, AND I

WANT AT THE SAME TIME TO SEND YOU THIS MESSAGE TO TELL

YOU HOW DELIGHTED I AM THAT YOU ARE GOING TO BE ABLE

TO SERVE.=

JOHN F KENNEDY.

THE COMPANY WILL APPRECIATE SUGGESTIONS FROM ITS PATRONS CONCERNING ITS SERVICE

Edward Teller and Garwin chat at the IBM Watson Scientific Laboratory at Columbia University. October 1960. *Photo courtesy of IBM.*

Garwin in doctoral gown at the University of Chicago. December 17, 1949. *Photo courtesy of Robert Garwin.*

Garwin's parents, Robert Garwin and Leona Schwartz Garwin. December 1949. *Photo courtesy of Richard L. Garwin.*

Garwin (left) with brother Edward L. Garwin. *Photo courtesy of Robert Garwin.*

Garwin's father, Robert Garwin, with electronic multimeter and vacuum tube tester. 1937. *Photo courtesy of Robert Garwin.*

Garwin (right) with Jerome B. Wiesner, science advisor to President John F. Kennedy. Garwin worked closely with Wiesner since 1953 on air defense, intelligence, and during Garwin's two terms on PSAC, 1962–1965 and 1969–1972, as well as all the years when Garwin was a consultant and panel member of PSAC. *Photo courtesy of the US government.*

Upper photo: National Academy of Sciences CISAC panel visit to Strategic Air Command, Offutt Air Force Base. Left to right: Evgeny P. Velikhov, Garwin, Sergey Rogov, and Wolfgang "Pief" Panofsky (in suspenders). *Lower photo:* Michael M. May and Garwin at Offutt Air Force Base. *Photo courtesy of the US government.*

Lois (second from left) with her and Garwin's children: Tom (far left), Laura (third from left), and Jeffrey (far right). August 1987. *Photo courtesy of Richard L. Garwin.*

Left to right: Ann Druyan, Garwin, and Carl Sagan in Moscow. April 1982. *Photo courtesy of Bernard Lown.*

Garwin (left, facing camera), Luis W. Alvarez (next to Garwin), Prof. Rene H. Miller (far right) of MIT, and Dr. Vincent V. McRae (slightly visible between Alvarez and Miller) of the White House Office of Science and Technology. McRae was the executive secretary of Garwin's PSAC Military Aircraft Panel and the Naval Warfare Panel. *Photo courtesy of the US government.*

Left to right: Paul Bracken, Jimmy Carter, Kurt Gottfried, and Garwin in Plains, Georgia. April 1985. *Photo courtesy of the Richard L. Garwin collection.*

Garwin (left) with his longtime IBM physicist colleague James L. Levine and the detector used in their search for gravity waves. 1972. *Photo courtesy of IBM.*

Ivy Mike mushroom cloud. *Photo courtesy of the Nuclear Weapons Archive.*

Hydrogen bomb. *Photo courtesy of the Nuclear Weapons Archive.*

Garwin receiving the Presidential Medal of Freedom from President Obama. November 22, 2016. Photo courtesy of the US government.

lives were indeed saved. The sensors were particularly useful at night, in the fog or rain. By using the detectors to find where the enemy was massing, and being able to shell from a distance, they avoided hand-to-hand infantry combat against overwhelming numbers that could have been disastrous.

Soon the navy was using them to protect the channel to Saigon harbor.[36] The DCPG kept improving the system. The use of the sensor spread throughout the battlefield but never had an important role in slowing down or winning the war. The JASONs never again did anything as dramatic as what happened at Khe Sanh; the fence was generally considered a failure, largely, many JASONs said, because the air force was not interested in letting machines—and indirectly civilians—tell it what to do.

"It was not an Air Force system," Garwin said about the wall. "The Air Force has an R&D establishment, they have doctrine, they have equipment and what not. They had no interest in making [JASON's plans] work."[37]

By this time, the war was becoming a huge social issue in America, and with the military preventing them from succeeding, many of the JASONs pulled out of the DCPG. Garwin did not. The JASONs were paid to deliver a report and deliver the report they did, the "Air-Supported Anti-Infiltration Barrier." It went to the sponsor that paid for it. That the sponsor was unwilling or unprepared to make full use of the report was not something one would quit over. Again, what is the role of the scientist advisor? *You asked, we answered. What you do with the answer is not our problem.*

Finally, the DCPG disbanded, its job done. McNamara's Wall may have been a failure outside of Khe Sanh, but the advantages of electronics did not go unnoticed.

One day, as the war grew in the public psyche, Garwin was flying out of LaGuardia to Boston on the Eastern Airlines shuttle. He was sitting by chance next to a young woman. She and her husband recognized him. Both were in the antiwar movement. When the plane landed and the passengers began filing out, the woman stood up and began screaming, "He is a baby killer!"[38]

Others in JASON and in the science establishment who had done work for the executive branch were running into the same thing, in some cases much worse, including constant picketing, heckling, and even physical threats. Garwin's home in Scarsdale, New York, was picketed, but he knew about it in advance and told the neighbors, who were sympathetic, just to stay inside. It ended peacefully.

Some of the JASONs were ashamed of the work the organization did in Vietnam. At a meeting of the American Physical Society in New York in 1973, Goldberger, then department chairman at Princeton, standing in a garish ballroom, quietly described his anguish: "I think we made a mistake. We misjudged the kind of people we were dealing with. Wise men would have known better but there were not many wise men in those days. I am not trying to walk away from my responsibility. You cannot rewrite the past."[39]

Others were torn. Sid Drell, at the same meeting, said, "If anyone has any confidence in one's government, one must do something. I've seen the government do things I like and things I dislike. We need to have critics not just on the outside, but on the inside too."[40]

For the first time in its history, the JASONs were a public target, in part because of references to the organization in the Pentagon Papers, which accurately described their war efforts. In many ways, the references in the documents published first by the *New York Times* and other newspapers were complimentary, pointing out that the wall was designed to reduce the bombing of the North and save civilian lives.

"Coming as they did from a highly prestigious and respected group of policy-supporting but independent-thinking scientists and scholars," the Pentagon Papers stated, "these views must have exercised a powerful influence on McNamara's thinking."[41] McNamara himself admitted this was one of the reasons he resigned.

Oddly, that part was usually ignored. At university campuses, students produced lists of JASONs, not always accurately, and JASONs continued to be subjected to harassment. Many, such as Goldberger and Drell, could hardly give public lectures without disruption, such as what happened at the New York APS meeting. Sometimes they would try to negotiate a deal, something like, "Let me give my talk and then we can discuss Vietnam later." Sometimes it worked. In 1972, physicists at Columbia University were barricaded in their office building for three days. Protestors demanded the faculty members of JASON either quit the organization or resign from the university. Garwin would do neither. Students confronted Garwin in his office at Columbia and at his home. He wrote in *Christianity and Crisis*:

> Remember the State of California [loyalty] Oath of 1950 and the long and degrading history of political tests for membership in a university. Which faculty member is sure that the years hence, 50 faculty members could not be found to demand either his own recantation or his dismissal? I long for one end to the Vietnam War. But lies and violence here at home and the attempted denial of legal rights of individuals (by antiwar activists or by others) will only injure further our society, which sorely needs all our energies turned toward improvement of its mechanisms and substance.[42]

One scientist's garage was set on fire. Offices were broken into. There were large demonstrations on various campuses, especially the University of California at Berkeley, sometimes led by young scientists. JASONs were labeled as war criminals. The demonstrations were not limited to the United States; JASON members found themselves under attack in Europe as well.

Drell, who had little to do with Vietnam, was also targeted. He spent almost all his time at PSAC, he said, working on missile survivability and missile defense. "I had absolutely no contact with the JASON study," he said. "That didn't matter. My name was on the list."[43]

In the spring of 1972, Drell was a visiting professor at the University of Rome. A grad student got up and said he knew Drell was a JASON and he would have to explain what he thought about the war and then there would be a vote on whether Drell could give his seminar. Drell said, "Like hell you will." It was fascism, "an inquisition," he told the student, and he would not participate. The student left and came back with a small mob. Drell walked out unharmed.

The next day a group of students followed him around campus hurling insults at him, "which I did not understand because they were in Italian." At Corsica, the school closed down because the students couldn't agree on whether Drell should give a lecture.[44]

Garwin found even social events were tainted. At one dinner party he was attacked by someone for what was perceived as Garwin's support for the war. Garwin said he'd be happy to discuss the war but warned he actually knew what he was talking about.[45] A pamphlet produced by one antiwar group depicted Garwin as ignoring the threat of strontium 90 from atomic tests in milk for children. The pamphlet said he spoke in favor of nuclear testing in the atmosphere and was quoted as saying, "Well, what's a few dead babies or mothers?" Garwin denied saying any such thing. In fact, he opposed atmospheric testing, and the quote is not credible. A rewritten version of the dinner party then had him comparing babies and mothers to the Jews in Germany.[46]

The physicists (and those under attack often were not just JASONs) fought back with little effect. "What is under attack," Garwin wrote in *Christianity and Crisis*, "is the right of an individual, in his own time, away from his regular job, to engage in a legal activity to which some individuals are opposed."[47]

CHAPTER TEN

SUPER SONIC TRANSPORT

By the mid-1950s, it was evident that the jet engine had revolutionized air travel, cutting destination times by half, making it possible to cross oceans without stopping. Propeller planes flew around 20,000 feet, still bumping along in weather, while jets could fly almost 20,000 feet higher, above the weather, and handle the trip across the Atlantic nonstop. And it seemed obvious that it was only the beginning. The Boeing 707, the Douglas DC-8, and the British de Havilland Comet were quickly replacing the old propeller-driven aircraft in commercial fleets. There was no reason to think that speed would be a limit. Military aircraft regularly flew faster.

In some ways the '50s and '60s were a golden age of engineering, perhaps reaching its inventive beginning with the ascendency of John F. Kennedy to the White House. It was the height of the Cold War, and he saw competition with the Soviet Union as a race, pushing engineers and scientists to exceed their dreams. They accomplished things later engineers could not, mostly because the United States government has retreated from these projects after the goal was accomplished. Work began to send men to the moon. The Apollo program would not only succeed in taking humans off Earth, but it triggered a scientific and technical revolution that is still ongoing, everything from how we use computers to how we monitor blood pressure. Work also began to make airliners go even faster. Because of the technological advances supersonic flight required, it was likely to initiate yet another explosion of technology.

Only America and the Soviet Union had the resources for the moon race, but the quest for supersonic flight was different. The British and French were, in some ways, ahead of the US in building a supersonic airliner. The Supersonic Transport (SST), however, became a political toy. Garwin was a crucial player.

The Englishman Frank Whittle was one of the inventors of the jet engine. Sir Geoffrey de Havilland put a jet airliner with four of those engines embedded in the wing, the Comet 1, into the air in 1949. Tragically, it suffered a series of catastrophic decompressions accidents and was finally grounded. Experts blamed design flaws in the fuselage. De Havilland solved the problem several years later with the Comet 4, but by that time, the Boeing 707 had launched the jet age, and Pan American World Airways was flying them across the Atlantic.

The French were in the jet race as well. Sud Aviation introduced the Caravelle, a small, beautiful, snub-nosed aircraft—with British engines. It set a precedent for British-French cooperation and, with a sale to United Airlines, showed that the French could sell to an American carrier. They were all subsonic.

French aviation pioneer Marcel Dassault was interested in the supersonic frontier. His Mystère IV-B jet fighter became one of the first planes to break the sound barrier in level flight in 1954, just months after an American F-100 did it. Four years later Avions Dassault's Mirage III-A flew at Mach 2.

Two companies, British Aircraft and Sud Aviation of France, began designing a supersonic airliner together with the help of the British engine company Bristol-Siddeley and France's SNECMA. It was to be called, appropriately, the Concorde. America had taken the lead with subsonic jets and was dominating European skies; the Concorde was going to jump ahead of it. The British and French signed an intergovernmental agreement in 1962 to share the cost, with the four companies working as contractors. The two countries saw it as a race, fueled at least in part by resentment that their two national air carriers, British Overseas Airways Corporation (BOAC) and Air France, were buying American jets.

To Kennedy, that was a challenge. Neither Boeing nor McDonnell Douglas had anything like it on design tables. It was clear that neither company could begin work without massive government aid.

Flying at supersonic speeds was technically more complicated than staying below the sound barrier; the physics were different. Moreover, the airlines were busy buying every jet the two companies could make. They were too preoccupied handling this revolution, and had little time to worry about the next one.

The head of the Federal Aviation Agency, Najeeb Halaby, however, thought commercial supersonic flight was inevitable, and in 1961 he got Congress to appropriate $11 million to begin a feasibility study.[1] As is usual in cases like this, the result was positive. If you want to build something you can always find a reason. He went to Kennedy with the report, and the president then, in true Washington fashion, formed another committee headed by Vice President Lyndon Johnson, a SST supporter. Things were gliding along nicely until Juan Trippe, the chairman of Pan Am, got into the act.

Pan Am had very few domestic routes, but it was America's designated international flag carrier, government subsidized in some routes, although unlike many other countries' flag carriers, it was publicly owned. If you worked for the US government, you flew Pan Am on international flights. If you were a country that wanted to land its planes at US airports, you made room for Pan Am at yours. It made its money—and lots of it—by taking passengers from large American cities to overseas destinations. Howard Hughes's Trans World Airlines was its only competition. Many of Pan Am's passengers were fed to Pan Am by domestic airlines, who had few if any international destinations. It even flew one route around the world, Pan Am Flight 1, perhaps the most famous airline route of any American carrier. Trippe was a major force in the industry, and his power was sufficient to influence the aircraft manufacturers. Boeing's 707 was largely inspired by Trippe, and later he would almost design the 747 for Boeing. He could not be ignored.

Kennedy was about to make a declaration of a project to build an American SST when Trippe announced that Pan Am was going to place a preliminary order for six Concordes, which he said would fly the Atlantic in two and a half hours. The contract wasn't binding, but

Trippe had signed an agreement to make sure Pan Am was covered should the Concorde fly before American competitors were ready. He had no intention of letting BOAC and Air France fly at supersonic speeds across the Atlantic while America's flagship airline was lumbering along in his 707s and DC-8s.

Kennedy was furious. He called Treasury Secretary C. Douglas Dillon on June 4, 1963. "Have you seen what Juan Trippe did? How could he do that when he knew we were about to go ahead," he raged to Dillon.[2] He thought the Pan Am order would undermine US efforts. Dillon said Trippe told Halaby that he was under pressure from the British and French governments.

"He's giving me the best argument for not having one airline represent the United States, I've ever heard," Kennedy said. "I'm going to spend our time here screwing Pan American. . . . I mean, didn't we have an understanding with him that he wouldn't go ahead when we were trying to come up with our proposal?" He told Dillon he should call Trippe and "and stick it right up his ass."[3]

"My God," he said, "I had it in my speech tomorrow."

Kennedy knew that the American SST would cost a huge amount of money, and if the Concorde was built before the US had a competitor, it would hugely affect America's balance of payments. The next day, in his speech to the graduating class of the Air Force Academy, he said: "It is my judgement that this Government should immediately commence a new program in partnership with private industry to develop at the earliest practical date the prototype of a commercially successful supersonic transport, superior to that being built in any other country in the world."[4]

He announced the National Supersonic Transport program. He then went to Congress and asked for $750 million—stipulating that was the maximum—and the manufacturers would have to put up 25 percent.[5]

The airlines were more or less ordered to buy the plane, if only so the government could get its money back. The carriers had to put up an initial, nonrefundable payment.[6]

The rationale for the economics underlying the SST was that airlines could double the number of flights they could fly in a day. If it only took two and a half hours to fly from New York to London, they could turn the plane around and send it back in enough time to do it again. That presupposed, of course, the price of fuel remained low. It would be so attractive they could charge a surcharge, especially for business fliers who had to cross the ocean or the continent in a hurry. The FAA guessed there would be a market for 500. Requests for proposals went out to Boeing, Lockheed, and North American Aviation for the body, and to Curtiss-Wright, General Electric, and Pratt & Whitney, who all built engines.

Boeing and North American came up with plans that resembled supersonic bombers they built for the US Air Force; Lockheed's looked like the Concorde design on steroids. Boeing and Lockheed were told to try again; North American dropped out. Both produced full-scale mock-ups. Boeing and GE won. They promised the planes would fly by 1970. Seven airlines ordered 122 planes, eventually including Pan Am.

The project was not universally welcomed, with environmentalists sounding the first alarm, warning the plane would damage the ozone layer protecting the Earth, and the noise from the sonic boom would be destructive. Liberal politicians, such as William Proxmire (D-Wisconsin), disliked the idea of federal subsidies. In 1971, the US Senate voted to cut off funding, and the House of Representatives followed. Labor Unions fought back, warning of a recession in the aviation industry.

Kennedy commissioned an outside review of the project just before he was assassinated. The review commission, headed by Eugene Black, former president of the World Bank, and Stanley Osborne, chairman of Olin Mathieson, finished the report and handed it in to Lyndon Johnson, then in office. It strongly stated the project was a really bad idea. As the controversy raged in Washington, Richard Nixon, now the president, called for multiple investigations into whether the US should build an SST.

Lee DuBridge, Nixon's science advisor, told the media when he took office that he was forming two scientific panels. One, under Goldberger, would look at ballistic missile defense, the other, under Garwin, at the SST. DuBridge assured the reporters the panels would report in a month and that he was looking forward to sharing the results with the press. No one else in the White House was interested in sharing the results with the press, Garwin said.[7] What Nixon really wanted was some prestigious committee to say yes, absolutely build it for American prestige, and that the potential for considerable profit was at stake.

Garwin had done some work on the SST for the White House when President Lyndon Johnson had asked Robert McNamara, secretary of defense, to weigh in on the government program to develop a commercial aircraft that would carry passengers at three times the speed of sound. It was clear to Garwin that the SST was a bad bet. In order to make the plane a paying operation, it had three requirements: it had to be safe, economical to operate, and environmentally sound. The planned plane, now dubbed the Boeing 2707, would satisfy none of those criteria.

"We said the . . . government should admit that it's not going to satisfy those goals or it ought to cancel the program," Garwin said.[8] This is typical Garwin: you can go ahead and build it if you must, but don't pretend it is a good idea.

"If it flew over land," he said, "the sonic boom would be unacceptable."[9] The noise of the engines would equal that of fifty subsonic 747s taking off, and there was no economical way to quiet the plane. Sound suppressors would work, but they would add weight to a plane that was already struggling to take off; the plane would not have enough range, and the sonic boom would annoy millions of people and probably cause property damage.

"You could not have a supersonic aircraft fly over land," he said.[10]

Congress would have to put up the money and knew about the panel, as did everyone else. The month went by, and nothing was released. The report was suppressed, not entirely for political reasons. The design of the plane had it capable of flying at almost Mach 3, three times the speed of sound and much faster than the Concorde. At that speed the temperatures on the fuselage would be so high from friction with the air that aluminum would melt. The Concorde didn't have that problem, so it was built with a mostly standard aluminum hull. But the 2707 needed a titanium hull because only titanium was light enough to fly and could resist the temperatures. The problem was that technology only existed on the air force's reconnaissance plane, the SR-71, and even the existence of the SR-71 was a secret. It also meant the Pentagon would get dragged into the project, and the Pentagon was unwilling.

Finally, on March 13, 1970, the *New York Times* printed a story with the headline, "Report on SST Said to Be Kept Secret." Congress decided to act and asked Garwin to testify three times. Each time he told them that a supersonic airliner was a waste of money. The planes could not pay for themselves, and would be an environmental disaster.[11]

Nixon had already approved of the scientists on the panels testifying about the antiballistic missile (ABM) system, and when asked about testimony on the SST he offhandedly said everyone should have the benefit of their expertise.

"He didn't really mean it," Garwin said.[12]

The issue of the sonic boom alone would have killed the project. Everything Garwin had in his report in 1969 was borne out by the thirty-year record of the Concorde. Essentially, there are two types of booms, one from the aircraft straight down to the ground as the plane breaks the sound barrier. It happens just then, is quick, and is totally unavoidable. In steady flight this is called a "carpet boom," an N-shaped field of sound as the plane travels at cruising altitude. A thermospheric sonic boom that oddly, doesn't show up until ten or fifteen minutes after the plane has passed through because in effect it travels upward from the plane and then is reflected by the atmosphere by wind and temperature layers. It lasts ten to fifteen seconds and then quiets. When the Concorde arrived at the JFK airport soon after it began service in 1975, the carpet boom could be picked up on sensitive microphones as far away as New Hampshire.[13]

In 1964, the FAA and the air force began a study, Operation Bongo, to see just how annoying or dangerous the booms were. It's possible people would just get used to them and pay no attention, advocates believed. So the air force sent supersonic F-104s over Oklahoma City, every day at specified times.

According to one report, many people did shrug them off or make use of them. One secretary, knowing the schedule, used them as an alarm clock. The boom rattled her windows, so it was time to get up. Seven a.m. At 7:20, another boom told her it was time to start her day. Construction workers learned to use the 11:00 a.m. boom to know when it was time for a coffee break.[14]

This went on for six months. Oklahoma City had been chosen because it was aviation-minded with an air force base and a FAA facility in town, somewhat biasing the sample. And the booms weren't as strong as those that might come from a SST taking off, and they were only in the daytime.

Nonetheless, 4,900 people filed damage claims, although mostly for minor destruction. Office towers reported 147 cracked windows.[15] Surveys at the end of the six-month period showed the one-quarter of the people in Oklahoma City said they could not live with the noise.

Another test out of Edwards Air Force Base in California, using an XB-70, a supersonic bomber about the size of the Boeing 2707, cracked windows. A transcontinental flight by the B-58 bomber, slightly smaller than the SST, lit up police switchboards across the country with reports of explosions and broken glass.

A survey by the Stanford Research Institute said:

> It is expected that about 65 million people in the United States could be exposed to an average of about ten sonic booms per day. . . . A boom will initially be equivalent in acceptability to the noise from a present-day four-engined turbofan jet at an altitude of about 200 feet during approach to landing, or at 500 feet with takeoff power, or the noise from a truck at maximum highway speed at a distance of about 30 feet.[16]

That ended any chance the plane would fly over American land.

The Garwin report was devastating and exactly what Richard Nixon did not want to see. He had put all the pressure he could to get a favorable nod from Garwin, even leaning on Thomas Watson Jr., CEO of IBM. At a cafeteria one day, Watson went up to Garwin and asked him about the SST study. It was, Garwin said, the only time in his entire career at the company that happened.[17]

The plan was to build two prototype aircraft that would be tested for 100 hours, seven of which would be at top speed of Mach 2.7 and at 60,000–70,000 feet altitude. Garwin testified he had no idea why they were planning on building two test planes when one would do.[18] Neither was going to do any science at that flight regime. The SR-71, no longer a secret, had flown faster and higher. It was not clear that the testing would involve oceanic flights, which would be the major route.

By this time, Garwin had congressional testimony down to a science. This was before laptops or even desktops, so he brought along a huge suitcase that he propped up on a chair next to him at the witness table. If there was a question he could not answer off the top of his head, he reached into the suitcase and pulled out a document.

"So I would have this big suitcase and I'd come into the Senate hearing room and pull up a chair next to me and open the suitcase. So if there was a question about something, I could pull out the document. And, of course, in those days I think in testifying you had to bring along a hundred copies of whatever it was you were going to present. So I was fortunate in working for IBM, even in the 1950s and 60s, we had the chain printers with fan-fold paper, and burster-trimmers. So I would bring home many pounds of such stuff and the children and my wife and I would go around the dining room table and collate these things and staple them so I could have a hundred copies," he said.[19]

He was one of the inventors of IBM's laser printers.

"There are few pleasant surprises investing of this type of prototype," he told a Senate committee on appropriations on August 28, 1976, "there are many unpleasant surprises."[20] Building the prototypes wouldn't give much information on how much production models would cost because they would be manufactured using different processes. The prototype SSTs would essentially be hand-built.

On the matter of the sonic boom, he said Boeing admitted the boom beneath the aircraft would register 2.5 pounds per square foot during cruise and 3.5 psf when the plane was accelerating. The tests would not tell Boeing any more about the sonic boom problem than it already knew, he said.

The planes could not be profitable if they were restricted to flying over water, so there would be immense pressure to change the rules to permit overland flights. Too much money was at stake. If the plane was outfitted with less powerful engines to hush the noise at takeoff, it would require longer runways, and only two airports in the country could take them. He didn't say which, but presumably one would be Kennedy in New York. And even then, it would be too noisy. And, he added, because of the limited range, it would make air traffic control more complicated. If weather caused landing delays, for instance, air traffic control would have to put the SST in front of the line before it ran out of fuel.

And the cost of proceeding would be in the billions, and private financing wouldn't do it. Further, there was no chance the government would get its money back.

"Enough information is available right now for the Senate to cancel the SST," he said. "The testing of the prototypes may provide additional negative information—it will not provide any more encouragement to the program. It is fundamental that if a program is to be terminated, it should be terminated as soon as possible." America would be better off building better train service, he testified.

The report torpedoed Halaby by suggesting only one plane be built for research purposes and that the whole project be taken out of the hands of the FAA on grounds it did not have the resources to run it.

It was safe to assume Garwin's testimony and the report did not go over well with many, particularly the unions who would be building the planes. But the manufacturers were happy because they actually never wanted to build them. They were convinced they would lose money on the deal and were displeased with the cost sharing. The airlines, which really did not want to spend the money either and had been essentially ordered to buy the plane, were pleased, as were environmentalists.

Garwin said he voted for Nixon in 1968. When he and Lois moved to Scarsdale, New York, in 1955, then a heavily Republican town, they decided that Lois would register as a Democrat, Garwin as a Republican. Still, he was impressed by Nixon's promise to end the Vietnam War, but felt Nixon and his administration tended to distort science advice when talking to Congress. Garwin knew his testimony would be controversial, in part because White House advisors did not go public with their disagreements with White House policy. Nixon thought Garwin and other opponents of the SST were "environmental fanatics."[21]

Even within PSAC, the testimony had been controversial, with many in the organization believing it inappropriate for a member of the committee. They believed Garwin should have simply resigned before testifying. He believed his resignation would not have done any good. There had been some precedent during the battle over the antiballistic missile issue, when Jerry Wiesner testified against the plan, possibly with Nixon's consent as related by Science Advisor Lee DuBridge. But Garwin also knew that PSAC was merely a ghost of itself, that the White House was full of Nixon aides who wanted to kill it, and even if his testimony led to its demise, little would be lost. His testimony would likely be the coup de grace.

The White House was furious. Nixon hated what he considered disloyalty. Here were people acting independently, going against the administration's plans.

Congress killed the SST in 1971. The Concorde flew for thirty years for British Airways (BOAC's successor), and Air France on limited routes highly subsidized by the two governments. A total of fourteen Concorde aircraft were built.

The irony is that in the twenty-first century air transportation is slower than it was a half century ago as the planes—all subsonic—are hindered by traffic constraints. And people no longer walk on the moon.

According to many, including his friends, it was Garwin's eventual testimony on the SST that doomed PSAC, but it may have only been the proverbial straw. His opposition to other Pentagon projects may have played a role. Relations with the White House had been deteriorating for years as was PSAC's influence. Nixon removing the members from the White House cafeteria was a sign. Also, Nixon and Lee DuBridge had reached the point where Nixon would talk about what a bad job DuBridge was doing behind his back while pretending to be respectful in front of the former Caltech president. DuBridge, who helped with the development of radar at MIT's Radiation Lab, never caught on. Nixon just hated firing people.[22]

Except for Eisenhower and Kennedy, relations between the president and PSAC were never close. Garwin said he met every president "a couple of times," usually photo ops, formal pictures taken in the Oval Office, sometimes at receptions, and, before Nixon threw them out, in the cafeteria.[23]

Things finally reached a head when the White House sat on the SST report, stonewalling not only the public, but Congress, which had to fund any program. Opponents of the SST, which included most environmentalists, tried to get copies, but that did not work. Both the FAA and the Department of Transportation kept feeding Congress information. Garwin said that Boeing and General Electric were telling the truth, but that the government agencies and departments were not.[24] PSAC held discussions on what to do. When Garwin was asked to testify, he wanted DuBridge to find out if he should resign from PSAC first, suggesting DuBridge take it up with Nixon. He said he was generally going to say what many in Congress already knew, that the program was ill-advised. Killian said he shouldn't testify, but word came back from DuBridge: Nixon did not object. Whether DuBridge was telling the truth, had misheard what he was told, or Nixon lied, no one knows.

MacDonald had already testified for the ABM system under pressure, Garwin said. Garwin went before Congress and has been accused of using information he had obtained as a member of PSAC in his testimony, and while it probably was not classified, it was supposed to be confidential. Nixon pulled the plug on PSAC. His excuse was that PSAC was costing $2 million a year and he had other places he wanted to spend the money.[25]

"Whether that was the right thing to do," Garwin said about his testimony, "I don't know." He had been careful to use only information already possessed by Congress, he said, adding his own judgment.[26]

There was another possibility for Nixon's antipathy to PSAC. All the telephones in the government, certainly in the executive branch, had stickers on them warning that all telephone calls were subject to monitoring. In the Nixon White House, they meant it. Several members of PSAC were overheard talking to Democratic Party headquarters about the upcoming election.[27]

CHAPTER ELEVEN

OFFENSE

The Germans' V-1 rocket was a cruise missile, cheap, inefficient but effective, as the citizens of London discovered. In German they are called *Vergeltungswaffe*, which literally translates to "vengeance weapon." It was unguided, unlike the later V-2, which was the first long-range ballistic missile. You could actually aim the V-2. The Germans in France simply pointed the V-1 at London and hoped it would land there and explode.

Cruise missiles use aerodynamic lift to fly, just like a plane does instead of the momentum from the blast of a rocket. It is continuously thrusting—the air-breathing engine is on until it runs out of fuel and gravity drops it to a target. Modern cruise missiles, aerial torpedoes, are much better and far more accurate and effective than the V-1. Those who launch cruise missiles know exactly where the missiles will come down, perhaps even which window they would fly through. They are far more expensive as well. The more fuel it carries (and the less explosive) the farther one can fly. Five thousand kilometers (3,100 miles) is entirely possible if fuel is substituted for some of the explosive in the warhead. For very long range, ballistic missiles are better, but for intermediate range, cruise missiles are superior. They are slower than ballistic missiles and easier to develop. Cruise missiles are like airplanes: you can test them with short-range flights and later use them for the long range. They fly at an altitude that maximizes their range, sometimes skimming the surface. Cruise missiles are easy to launch, easy to fuel. They fly on kerosene, not rocket fuel; they burn oxygen from the air.[1]

In the 1970s, Garwin recommended cruise missiles based in Europe instead of having US air bases spread around the continent, which again made him directly in opposition to the US Air Force. The accuracy of cruise missiles depended on the guidance system being used. Cruise missiles could be fired from the ships that brought them to Europe or from depots. Being able to open a crate, Garwin said, lift the missile up and fire, and keep firing repeatedly would be better than launching an attack from air bases with either bombers or missiles in silos. But air force officers wanted to command men, not missiles. They were somewhat cheered up in 1978, when tests of Tomahawk cruise missiles fired from submarines (in a demonstration with press from around the world watching) burst through the surface of the water, roared into the air, and fell ignobly back down into the water.[2]

Still, Garwin said, "You couldn't deny they existed and were effective, but the Air Force didn't like them. . . .[3]

"They didn't get started in the modern age until my naval warfare panel espoused long range cruise missiles—nuclear armed and non-nuclear armed—in the 1960s." The liaison for the panel with the Pentagon was Admiral Elmo Zumwalt, one of the Pentagon's stars, which made things easier. Zumwalt was in favor of naval missiles. But Zumwalt was from the US Navy. A joint cruise missile development office was established in early 1977 by William J. Perry, Jimmy Carter's undersecretary of defense for research and engineering, in order to get the air force to take seriously the capabilities of the cruise missile. Air force wanted the B-1 bomber and wouldn't accept that long-range cruise missiles could be built for the job.

Garwin thought he had a better idea than the B-1: take a Boeing 747, then the largest commercial airliner in production, modify the fuselage, and turn it into a triple threat. One day it would be a tanker, being able to refuel warplanes in the sky, the next day it could be a cargo plane (and already was for many of the world's airlines), and on another day, it could carry and launch cruise missiles. Once you modified the tail, cruise missiles on rollers could be shoved out the back of the plane using parachutes. When the plane got farther ahead, the missile would ignite and follow its navigation directions to the target. Another way was with a device like a revolver. It would shoot a missile out the back of the plane, rotate and fire another, and keep going. Or, the rotary launcher could be lowered out the bomb bay of a plane, or there could be slots on the side of the plane where the missiles could be shot out from. The missiles also could be launched out of a launch tube slightly larger than a window that could potentially fire ten cruise missiles a minute.

The air force would have none of it. At one time an air force general went to a JASON meeting, possibly to undercut Garwin, and said the B-1 couldn't carry a cruise missile with the right range and nuclear warhead. They needed a penetrating bomber to do the deed. Garwin pointed out that the navy had carried just such a test on a smaller aircraft, an A-6, that same day.

"That was the story he had been wound up to tell," Garwin said of the general.[4]

On April 17, 1975, Garwin testified before the US House Armed Services Committee, reinforcing his reputation as a gadfly, and it got personal. The air force testified that it had matched Garwin's idea against its plan for the B-1, and the B-1 was a better idea.

"I cannot agree with the proposed funding for the B-1 bomber program," Garwin testified. "I have followed the B-1 from its inception. The B-1 was hardly put into valid competition with alternative programs at the time of its transition to full-scale system development, and continued expenditures on the B-1 will indeed buy us the shade rather than the substance of first-class military power."[5]

The fleet of B-1s would cost $80 billion plus operating costs once it was built, he testified. The only purpose they would serve would be the "delivery of nuclear weapons in strategic war." That was a lot of money for a warplane that no one may ever need.

An air force general countered, calling Garwin's modified 747 a hypothetical plane. No, Garwin said, the B-1 is the hypothetical plane; the 747 is already in production.

It got to the point where the air force had refused to testify at the same hearing if Garwin was allowed to ask questions. So Garwin asked one of the congressmen if it wouldn't be useful if he asked the air force general how long it would take them to launch forty cruise missiles from the equipment they had analyzed.

"An hour," the general said.

The navy could launch a missile in fifteen seconds from a submarine and could launch sixteen missiles from a sub in four minutes, Garwin pointed out. And they could do that even though the sub had to have its ballast adjusted after every shot.

Garwin said the air force had obtained a written agreement from the secretary of defense that no testing of a 747 conversion be authorized. That information was being withheld from the committee at the air force's request.

Major General Robert Lukeman, testifying for the air force denied this. "I feel an obligation to state for the record that I am not aware of and have never heard of any agreement between the Department of the Air Force and the Office of the Secretary of Defense regarding not doing something with the 747 aircraft. I want to say I have never heard of it. I'm not aware of it. I think this is important," he said.

"I will supply it," Garwin shot back. He did.

"I testified against the B-1 bomber," he said later, "because I wanted those nuclear weapons to be delivered by a cruise missile carrier." Part of the three-legged system of air, sea, and land, called the Triad, he supported. Modifying a 747 would be fast and cheap. All that was required was to change the interior to convert it from one use to another and if necessary add an aft door. The plane would carry cruise missiles with a range of 1,500 nautical miles that would fly at an altitude of just under 200 feet at almost the speed of sound, below enemy radar, making them very tough targets to find and hit. Moreover, Garwin said, it would be done a lot quicker than building the B-1 or modifying the B-52.[6]

"It wouldn't have to develop a new airplane," he said later. "Since the air force was all about developing a new airplane, they didn't like it."[7]

The air force brought in slides and a seven-minute movie to demonstrate that the B-1 was really cost effective.

During the testimony a congressman asked if Garwin had ever

been in the military, apparently trying to challenge his patriotism or his expertise. No, Garwin said, but he had been involved in defense technology for years. He didn't mention the hydrogen bomb. Later the congressman apologized to Garwin for the implication.

Naturally, Garwin lost. The B-1 was built, Congress members were not prepared to vote against expensive weapons programs, particularly if some of the new weapons were to be built by their constituents. Furthermore, there was a recession in the aerospace industry and the B-1 meant jobs all over the country, as well as taxes.

The B-1 has gone through various interruptions and modifications. It was intended to replace the now half-century-old B-52 bombers but has not. Now used as a conventional bomber, the B-1B, the current model, was in action in Kosovo, Iraq, and Afghanistan. The sixty-two-plane fleet is now undergoing a massive overhaul with a total price of $918 million.[8]

Again, Garwin opposed spending time and money on a weapons system he did not think was necessary, that could be replaced by something simpler and cheaper.

"I do that," he said.[9]

Meanwhile, in 1979, Garwin was appointed a full professor at the Kennedy School at Harvard, a difficult situation because he was still working for IBM and had duties there. It failed, he said.

"I decided that I could either do this thinking and having influence in Washington and writing about it, or that and teach, and it was the teaching and being away from my IBM colleagues that tipped the balance. In any case, I resigned my professorship at Harvard in 1981."[10] He went back to the IBM-government job.

The controversy over the Missile-Experimental, or MX, was a great example of what happens when a government agency, particularly an armed force, decides that size matters. It's also a classic instance of Garwin battling an entrenched bureaucracy.

Part of the passion for a new missile was that while ground-based missiles were vulnerable to attack even in their steel-reinforced silos, missiles onboard submarines were not. They were hard to find and destroy as they roamed the seas, and could be fired from offshore, drastically shortening flight time and the ability of the victims to retaliate. But the navy ran the submarines, not the air force, and inter-agency competition was fierce.

At the time, said Garwin's son, Tom, who worked in the Defense Department and Congress, the inter-service rivalry was particularly savage and often irrational, and even within the services, one branch battled another. The aircraft carrier officers in the navy wanted the defense of carriers to be entirely the responsibility of those ships. They did not want submariners involved.[11]

Another problem was the moral issue of having the missiles aimed at essentially civilian targets—cities—and that bothered just about everyone. And the missiles were not terribly accurate. Many felt it would be better to have the missiles aimed at the enemy's missiles.

"Here we had been successfully negotiating with the Soviets to limit defense against ballistic missiles because everything we had seen in the years before 1972 was that defense would be ineffective," Garwin said, "but the presence of even an ineffective defense meant that the people who had built it or had paid for building it would have to say it was effective, and the other side would see that it was ineffective, and could go around it or over it or through it. Every year we studied this. Two days every month the Strategic Military Panel of the PSAC met, and our recommendation was that the project shouldn't be deployed because it wouldn't work. From the point of view of arms control it would be better if there weren't defenses because even the whiff of a Soviet defense in the future made us build many more missile war-heads. It was highly destabilizing."[12]

The Russians had big missiles, so the air force, aided by the defense intellectuals (who generally would now be called "neoconservatives" or neocons), wanted big missiles, land-based and generally invulner-able. Missiles were cheaper than a bomber-led deterrent.

As for the MX, "You can't just say, 'I want it,' and stamp your feet," Garwin said. "You have to have some argument."[13]

There had to be a rationalization, so Paul Nitze, a former deputy secretary of defense, and others argued that the Minuteman missiles in their silos were too vulnerable to nuclear attack from Russia, and that this vulnerability could provoke a Soviet nuclear strike.[14]

The Russians had a missile, the SS-9, that had three very powerful nuclear warheads. That meant that one of their warheads could destroy three American warheads since the US had three in each Minuteman. One Soviet missile with three warheads, it was said, could destroy nine of America's. But given the vicissitudes of missile attacks, the Soviet threat was highly overrated. Their missiles were not that accurate, and weather and mechanical difficulties mitigated much of the threat, Garwin and the others felt. Moreover, the Minutemen were only one-third of the US's potential responses to an attack. There were missiles with warheads on submarines, and bombers carried bombs, all of which could obliterate the Soviet Union. Also, it was unlikely that any Soviet attack would destroy the entire US land-based missile fleet. The argument was similar to one about whether to build a hydrogen bomb. It didn't matter if the Russians got there first with a bigger bomb, since America had enough atomic bombs to destroy Russia and had enough missiles as well and the Russians knew that. That logic, however, did not impress what Eisenhower called the "military-industrial complex"[15] or its supporters of the armed forces. The SS-9's successor, the SS-18, ultimately was fitted with ten warheads and a very large number of "decoys" or penetration aids, so that it did indeed become a formidable attacking force.

Nonetheless, since the Minuteman system was one of the strongest pegs in the ability of the US to retaliate against a Soviet attack, its vulnerability would be a bad thing. Hence, the neocons and the air force said the US needed a missile system that was even bigger.[16] Putting a bigger missile into the same silos would evidently increase the problem of missile vulnerability, not reduce it. The reason for a bigger missile was simply that it was the biggest one that was allowed to be built under the arms control agreement, and that was justification enough for many. It was an example of the rueful observation in the arms control community, "the ceiling becomes the floor."

Around 1980, backers came up with the MX (ironically named the Peacekeeper then the Peacemaker). It was going to be a ten-warhead missile, the biggest allowed by the Arms Limitation Treaty, 100 tons. There was no hope of getting authorization, however, unless it could be proven that the MX was sufficiently invulnerable to Soviet attack, and that the new project, which everyone knew would be hair-raisingly expensive, would solve the problem. Moreover, if the MX proved to be vulnerable, it would make the potential Soviet threat worse.

Should the US build it? And if so, where to put it? Would the system prove to be provocative on one hand, and would it really be invulnerable? When the plans became public, the Defense Department called for studies. "Everybody and his brother studied it," Garwin said.[17]

A number of "basing modes" emerged in the missile battle. No one will accuse the designers of a lack of imagination.

At first the air force wanted to put the MX in the Minuteman silos, which Garwin pointed out meant substituting Minuteman vulnerabilities for MX vulnerabilities.[18] One option was for multiple protective shelters (MPS). There would be 400–500 missiles, 5,000 warheads, and 5,000 shelters. The missiles would be carried among the silos by huge vehicles, traveling seemingly at random. It would either unload a missile into the shelter or pretend to unload a missile then drive away. Even a spy watching from a distance might not know if there was a MX in the silo or not, and the Soviets would have had to knock out 5,000 silos to ensure total annihilation of the US threat. The best a single Soviet missile would do was to destroy one out of five thousand of the threat.

"It sounded funny then too," Garwin said. "How do you pretend? If you offloaded a 100-ton missile and drove away anyone can tell if the carrier was empty by the way it bounced. Maybe you should have a 100-ton dummy."[19] There were other ways for the Soviets to spot the ringers, using X-ray technology or detecting radiation from the real warheads. Among the other ideas was a series of tunnels in which the missiles would be buried, the so-called "racetrack." Or they would constantly be in motion so the Soviets couldn't possibly know where they were in the open countryside.

The missiles would be parked anywhere in the racetrack, and even espionage would prove difficult because it would take thirty minutes for a missile to get from Russia to the US, and in that time the missiles could be moved. When a Soviet missile arrived, the MX would be someplace else where it could launch. That scheme subsided quickly when people in Utah and Nevada, where the racetrack was to be built, objected strenuously in part because it meant the federal government would have to take too much public land. Their objections were probably moved along by Garwin and Drell who both traveled the area to convince the populace it was a bad idea. "I concluded that the MX was not needed," Drell said. "Why were we doing it? It had gotten very far along. I also concluded [a] couple of scientists were not going to be able to reverse that, but [at least there was] the idea of showing there was no sensible basing scheme."[20]

Paying their own way on separate trips, they went on what resembled a political tour.[21] They even helped the elders of the Church of Jesus Christ and the Latter-day Saints come out in opposition. As they traveled, air force officers followed them. By agreement the air force people never appeared on the same platform as Drell and Garwin, but would speak either before or after the two to counter what was said. Garwin and Drell found many in the area receptive.[22] The people in these remote rural and desert areas were, among other things, afraid that their place in the world would become a hotbed of espionage, with Russian spies pouring in to determine which of the silos had real missiles in them and which were decoys, followed by the FBI or someone trying to catch them. Drell said, "They were afraid it would turn into a police state."[23]

The MX sites would have attracted attention. As the system stood, the Soviets could destroy ten American warheads with two of their own, almost inviting an attack, Drell said. "It's very critical not to have vulnerable bases. We hit on this idea of small submarines and we pushed it as far as we could."[24] In a JASON report, Garwin and Drell said the only logical place to store MX was underwater, on subs. Freeman Dyson was a coauthor.

"Sid and I were so incensed by bureaucratic imperatives and national security nonsense promulgated in regard to MX missiles," Garwin said. "And we were driven to invent what we really thought was objectively the best deployment mode if we were indeed to have a ten-warhead MX missile—the SUM [Small Undersea Mobile] system." That, however, meant the navy, not the air force would launch the missiles.[25]

Meanwhile, ideas kept sprouting.

Garwin found himself on the opposite side of the basing dispute from William Perry. "We were both wrong," Perry later said.[26]

There were plans for diesel-powered airplanes that would stay aloft for two weeks at a time—Garwin said it was overkill, because you did not need to have them up that long. Even a day would do it. There was another scheme to simply drop the missiles out of the backs of cargo planes with small parachutes for orientation. They would shoot out horizontally then turn to 45 degrees vertically and blast off.[27]

Then there was fratricide.

Silos were built sturdy. US silos could withstand 2,000 pounds per square inch of pressure, 150 atmospheres. They could not withstand direct hits, but the incoming Soviet missiles were believed to be inaccurate and reliable enough to destroy one silo per warhead. If they hit one it would be by luck. If two warheads came in to destroy two adjacent silos, one explosion, with its radiation, debris, blast, would destroy the other warhead before it could be detonated—fratricide.

(In the original Minuteman missile silo deployment, spacing was chosen so that even aimed at a point halfway between two adjacent silos, one missile could destroy neither; therefore not more than one silo could be destroyed per warhead under the best of circumstances.)

Then there was the notion of putting super-hard silos on the south side of mountains. The theory went since Soviet missiles would probably be coming in from the north, they would explode on the wrong side of the mountain. That is generally foolproof—unless you send a warhead powerful enough to take out the south side of the mountain as well as the north. Of course, the Soviets also could send their ICBMs "the long way around" so that they would be what was known as "fractional-orbit bombardment systems." This would limit their payload to about one-third of the normal ICBM trajectory and impair the accuracy.

It also was pointed out that a nuclear explosion was seriously messy, with millions of tons of debris flying all over the place. Tom Garwin even coauthored a paper with John Steinbruner suggesting patterns the Soviets might fire to avoid missiles from having to fly through the debris of previous explosions. You simply fire the ones aimed at the farthest silos first. His father proposed what he called a "bed of nails" defense—sticking reinforcing rods up in the air to block an attack. The attacking missiles would be coming in on a straight line and a low angle to the ground. Because of the way the silos would be built, any attacking missile would have to hit the ground very close to knock out the missile in the silo. If you built a fence that was close in space, twenty or thirty feet high, you'd prevent those warheads from being able to explode within the distance to demolish the Minuteman silo. It wasn't the ultimate solution, he said, but no one ever thought of it before. Another idea was an explosive buried in gravel that would be detonated just as the enemy missiles arrived, producing a wall of flying gravel to destroy the missile or knock it off course. Tom said his father just came up with those almost instantly: "He is so inventive. Even though I was immersed in this extensively, I hadn't thought of these approaches."[28]

The Defense Department itself then came up with Densepack, which took advantage of fratricide: put very tough silos so close together they could not be destroyed more than one at a time. Drell called it "dunce pack."[29] If one of the attacking warheads destroyed a silo, it would prevent another missile from coming in because of the debris cloud. Garwin pointed out that all the Soviets had to do was adjust the timing to be simultaneous. That is, the arrival time of the missiles would be so precisely controlled that all would be within lethal range of their targets at the same instant so that no physical phenomenon from one explosion could arrive at the adjacent warhead before it had detonated itself. All the silos would be destroyed.[30]

The Densepack idea immediately ran into trouble in the Pentagon. According to Ted Postol, who was consulting at the Office of Technology Assessment with Drell and Garwin at the time, Admiral James Watkins, chief of naval operations, led a rebellion with three of the five Joint Chiefs opposing deployment. One of the reasons was intelligence that the Soviet Union was working on a missile with a tungsten tip and a thickened heat shield that would get through the dense defenses, and that the Densepack could not launch during the time it was being attacked. Postol did a study showing that if a Soviet submarine off the California coast fired a missile every thirty to forty seconds, it would pin down the MXs long enough for long-range ICBMs to arrive from the Soviet Union—with tungsten-tipped warheads.[31]

Another scheme, called Launch into Orbit, would place thousands of warheads in space over the Soviet Union if war seemed imminent. If the Soviets did start a war, the warheads would be commanded toward the ground. Postol pointed out that no one figured out how to get rid of them if war did *not* break out.[32]

"The controversy really got out of hand," Garwin said, and sometimes it became personal.[33] News reports on *60 Minutes* and in the *Los Angeles Times* even reported mismanagement on a grand scale, with some missile parts sourced from a local Radio Shack store. Garwin, who took this dispute seriously, was not amused.

The Minuteman already existed and was paid for. It was getting old, true, but it was upgraded with three warheads that turned out to be more accurate than people thought they would be. In all, a "very fine system," Garwin said.[34] But resistance seemed futile. The air force wanted new missiles, and the companies feeding off the Pentagon's teat wanted to build them.

Garwin was a member of PSAC's military panel. Henry Kissinger, Nixon's national security advisor, also hated PSAC. Partly it was because he had no control over what they did or said and partly because the scientists of PSAC knew about some things Kissinger did not, and Kissinger hated that. Nixon's staff resented their independence. Nonetheless, Garwin and the other advisors, under the leadership of Paul Doty, a fellow professor of Kissinger's at Harvard, and a close personal friend of his, met with Kissinger monthly in the White House situation room, and Kissinger would ask questions and give them something to study. It was highly classified.

"We were arguing for the elimination of missile defenses—all missile defenses," Garwin said.[35]

The existence of the military panel was completely unknown outside the White House until the flamboyant Italian TV journalist Oriana Fallaci, who had apparently mesmerized Kissinger, accompanied him into the situation room with a video crew and reported about it, even showing the color coding of the classified documents.

Garwin and Drell did get a change in the operations of MX, using GPS satellites, which he had a hand in developing. They would guide the missiles fired from submarines positioned a few hundred kilometers off the US coast, a technology that was employed.[36]

Garwin had been a member of the panel, which since 1957 or so had looked into the question of vulnerability and the penetration ability of missiles on both sides.[37] Starting in the mid-1970s, he was in the middle of the debate about what to do, first in secret, then in public, testifying before Congress, sending a torrent of mail, giving media interviews, and generally annoying the defense intellectuals.

In 1976, Garwin issued—"self-propelling"—a report, "Comments," in which he admitted that sooner or later "one will have to choose some missile and basing system, calculating the costs for a given survivable capability against a reactive Soviet attack by perhaps 1990 or 1995. The purpose of considering MX is of course to provide for the national security, which is impaired by diverting money from needed military systems to unneeded military systems."[38]

He suggested a force of 10,000 single-warhead, ten-ton ICBMs deployed in small military bases, even those in cities. They would each have a fixed target, say Moscow, launched by encrypted radio signals and retargeted, if necessary, either manually or by secure communications. The silos would be relatively small. He acknowledged that even if it made sense, Washington doesn't work that way and the bureaucracy would push for bigger silos. Also, the Soviet Union could just do the same. "Worse situations have occurred," he wrote. The US would still have a survivable deterrent.

Garwin also suggested in the report four states of launch-on-warning:

- Whatever is in operation now.
- Launching all the missiles armed with no means of disarming.
- A system in which a fraction of the rockets would be launched unarmed but after reaching apogee, about fifteen minutes later, could be armed with an encrypted signal.
- A system in which a fraction of the rockets could be launched armed and only disarmed with the receipt of a signal.

If the Russians tried to jam the signals when the missiles were launched in the third state, they would "effectively have destroyed their own country. It's pretty hard to keep them from doing that if they really want to," he wrote, and added it was not our problem.

He then went into the technology of communicating with the Minuteman and protecting the silos in minute detail, even discussing the cost, his point being that building the MX was unnecessary—just protect the Minuteman system.

When the basing controversy became public during the administration of President Jimmy Carter, Congress had formed the Office of Technology Assessment. It did a major study (MX Missile Basing) involving Sid Drell, among others, on its advisory panel. In the 335-page report published in September 1981, Congress looked at eleven possible basing scenarios. The basing staff included Ash Carter, currently secretary of defense, Ted Postol, a professor at MIT later to become a major critic of how the government spends its money on defense, and Robin Staffin, Garwin's research assistant at Harvard. They were not asked to judge whether the MX program was needed, only how to base them. They didn't choose one as a recommendation—again, they were not asked to do so, only to assess each possibility. They found five plans were feasible:

- MPS.
- Attacking first, before the Soviets could launch.
- MPS protected by an antiballistic missile system.
- Small submarines with the missiles launched from the sides.
- Dropping them out the back of aircraft.

They seemed to like the MPS best.[39]

In early 1983, President Regan asked Lieutenant General Brent Scowcroft to impanel a study on America's missile force. They came back with the suggestion that the US should rely on small, single warhead, survivably based.[40] Garwin, in a letter to the editor published in the *New York Times*, said there was "no suitable land-basing mode for a multi-warhead MX missile," particularly attacking dense-pack schemes, which had become the air force's favorite.[41]

Eventually fifty MX missiles were built and deployed in Minuteman silos. Then the Soviet Union collapsed and the whole scheme was canceled.

Now 450 Minuteman missiles sit in silos on four bases around the country, mostly in the West and Midwest. Each missile has only one warhead, as required by the disarmament agreements.

CHAPTER TWELVE

THE GREAT GAP

n the case of one missile defense plan promoted in the Nixon administration called Safeguard, space satellites—some of which Garwin helped develop—would detect the launch of an enemy missile. Radar in the northern climes would then predict the missiles' course. Safeguard involved two types of nuclear-armed interceptor rockets, Spartan and Sprint, along with infrared to spot a launch and determine the initial course, but used ABM radar in the terminal area to track the individual warheads to be destroyed. The five-megaton Spartan warheads would be used against clusters of warheads, and the kiloton-range Sprint would be used against individually tracked warheads.[1]

The trajectory could be altered by adjusting the payload. If the attackers used less than a full payload, for instance, the trajectory would be higher and the descent to the target steeper, making them harder to knock out. The attacker could also make the trajectory lower, cutting the time of flight to something like three minutes. The United States would need half a dozen powerful radar stations in places like Greenland and Alaska to do enemy missile tracking, and each station would be connected to "farms" of Spartans. Sprints would provide protection for the Spartans and for the radar stations.

But the effectiveness of Safeguard depended totally on the functioning of both of the two massive radar systems—one radar to observe the incoming warheads in space, and the other for battle management and discrimination of actual warheads from decoys. One way for an enemy to get around this, critics said, would be to equip the warhead of the assaulting rockets with decoys to confuse the radar. Decoys could be effective and are common in these weapons, but they are expensive because they have to look exactly like a reentry vehicle (RV). In a typical Garwinian suggestion, how about a spherical inflated Mylar balloon with aluminum foil skin? The RV would be enclosed in the balloon. Batteries inside the balloon would give an infrared signal similar to the one of an RV. In an attack no one could tell what was in the balloon, a warhead or nothing.

Another way of attacking the US would be bomblets of biological warfare material such as anthrax. The bomblets would spread out as they headlined downward, making them virtually impossible to stop.

In the end, Safeguard congressional support of the plan was argued to lend power to the American bargaining position in negotiations with the Soviets. It was funded after a telegram arrived from Gerard Smith, in Moscow as chief negotiator of the Strategic Arms Limitation Treaty (SALT I), that said if Congress did not provide authorization and appropriate funding, the US negotiating posture would be undercut.

The final bill for Safeguard was $10 billion.

The concept of shooting down ballistic missiles before they could do any damage goes back to World War II and the German V-2 rockets, which were aimed at British targets. Bell Labs, at the time, concluded it couldn't be done. The V-2 was too fast for any existing weaponry to immobilize, and that situation lasted until the development of high-speed computers. Then it was possible to match interceptor to attacker. But that didn't mean it was possible to erect an effective defense.

Work on an American ABM system was divided between the US Army and the US Air Force. The air force was in charge of some of the sensors, while the army was in charge of the ground-based radar and the interceptors, the rockets that were supposed to take out incoming attack missiles. The interceptors had nuclear warheads. It looked, said Greg Canavan, a senior science advisor at Los Alamos, who was involved in the subsequent controversy, like a "real system."[2]

Even with technological advances making this somewhat more feasible, opposition for the project came from Garwin and, this time,

Hans Bethe. Bethe's argument was that when the nuclear warhead of the interceptor went off, the fireball would blind the other interceptors. Bethe's opposition came early. Garwin's opposition came somewhat later.

The rocket worked fine, Canavan said. They had been tested hundreds of times. The nuclear warheads had been tested in Alaska and were not an issue. Thanks to the design work of John Foster at Los Alamos, the warheads had achieved the highest radiation temperature ever recorded in a nuclear weapon. The problem was the radar. Between 1969 and 1972, a series of meetings were held every few months to see if an effective radar system could be developed that would properly aim the interceptors at incoming missiles. Teller, of course, was involved, and Canavan said he was thoughtful about the technical problems. The meetings labored over equations, measuring things like how far the radar could see, wavelength of the transmissions, how much power was needed, etc. In many cases, it wasn't necessary for Canavan to reject proposals and papers from the participants. His usual practice to reject a proposal he thought wrong: "I just had to correct their algebra and they would go away."[3]

Canavan stated, "Those of us on the advocacy team said you could make it work but it would be such a kludge." The attacker could use that to its advantage, setting off a high-altitude nuclear blast rendering the following missiles invisible. Another complication was that for a system to work it had to destroy attacking missiles in the booster phase, while it was still rising, because once they reached their apogee, they could deploy clouds of decoys, and instead of killing the missiles "wholesale, we'd have to kill them retail," Canavan added. Each incoming rocket would be more expensive to destroy.

It was gradually concluded that the whole system was too complicated and prone to failure. "You don't want to bet your survival on it," Canavan said.

The time wasn't a complete waste. The army kept work on the interceptors, and the US may have a defense against attacks from villains like North Korea established from that technology. Called the Ground-Based

Midcourse Defense-Related System, and based in Alaska, it would not be a defense against a massive assault from Russia or China, Canavan said. Both countries could defeat the system by overwhelming it with numbers. A considerable amount of money was spent, "mostly for Viewgraphs," Canavan explained. Finally, President Ford killed the ABM, and it disappeared in the Anti-ABM Treaty. The good thing the work produced was that it convinced the Russians the US knew how to defend against the missiles, and for a year and a half the Russians thought the US was about to deploy the system, while in reality, the Nixon administration was desperately looking for one. Only a handful of people, including Garwin and Teller, knew no complete system existed.

On May 26, 1972, the US and the Soviet Union signed an Anti-Ballistic Missile Treaty barring either nation from setting up a nationwide ballistic missile defense system. Each side could set up two, one of them around their national capital, and the other an offensive missile site. The US selected as its site an interceptor facility in Grand Forks, North Dakota. In 1974, the number of missiles allowed were halved. The US eventually shut down its North Dakota defense because the cost greatly outweighed the protection it gave. In December 2001, President George W. Bush withdrew the US from the ABM treaty, saying it did not defend against attacks from "rogue" states like North Korea and that US and Russia really didn't need to base their relationship on their ability to destroy each other.[4]

For all the attention it received it might well have been one of the great secrets of the Cold War: for about twenty-five years, between around 1968 and 1993, America had no defense against a massive nuclear weapon from Russia. None. The US is still largely unprotected from such an assault. It could launch a counterattack that would do grave damage to Russia, but the devastation in the US homeland would be colossal. The United States of America could fairly well cease to exist.

Garwin doesn't think it was much of a secret. "I was telling everybody," he said. "It was more that the public resisted the truth. We didn't have any missile defense at all."[5]

The Soviet Union probably knew the US was defenseless because it was just as vulnerable. The US probably could survive even a massive attack long enough to destroy the Soviets, but the homelands of both countries would be rendered desolate and there would be no winners.

What has prevented—or at least fearfully discouraged—nuclear war is an irrational, somewhat juvenile situation called mutually assured destruction, or MAD. Even Garwin, the rationalist, agrees MAD is the reason we are still alive.

Another unpleasant fact generally ignored is that while there are some antimissile defenses against a nuclear attack from a rogue state like North Korea or China, something will likely get through. In July 2016, the Defense Department said it would install an antimissile system called THAAD—Terminal High Altitude Area Defense—in South Korea to defend against the North, despite objections from China, which says it does not want new weapons systems installed in the area.[6] It's still a MAD world.

Since the explosion at Eniwetok in the Marshall Islands, Garwin has spent much of his life working on disarmament, preventing the device he designed or its cousins from actually being used. Very few people know the threat these weapons pose better than he. He is now more afraid of terrorists than attacks from a nation-state.

Garwin had been working on ABM systems since his time at Los Alamos and since working on Lamp Light in his early days at IBM. Lamp Light was concerned with attacks by Soviet bombers.

"Part way through, I asked the management, 'By the time anything we recommend is implemented, the Soviets will be using intercontinental ballistic missiles (ICBMs) to carry their nuclear warheads to the United States,'" he wrote in an article with Bethe, a newly minted Nobel laureate, in a 1968 issue of *Scientific American*. "'What will we do then?'"[7]

Jerrold Zacharias said, in effect, "All in good time. We will solve that problem after we solve the air defense problem."[8]

Garwin continued, "Then serving as a consultant to the President's Science Advisory Committee (PSAC) from 1955 through 1973 or so, I was intimately involved in both analyzing and planning for ABM for the United States, and in penetrating actual and potential Soviet defense against US ballistic missiles and aircraft.[9]

"The primary fact is that the US and the USSR can annihilate each other as viable civilizations within a day or perhaps within an hour. . . . Each can at will inflict on the other more than 120 million immediate deaths, to which must be added deaths that will be caused by fire, fallout, disease, and starvation. In addition, more than 75 percent of the productive capacity of each country would be destroyed, regardless of who strikes first."

When Garwin wrote that, he added that the US had more than 2,000 thermonuclear weapons with an average yield of one megaton, and that former defense secretary McNamara once said the US needed fewer than 400 such weapons that could kill a third of the Soviets' population and three-fourths of its industry. Presumably, the Soviets could inflict similar damage on the US.

So it's not true, Dick Garwin said, that we have nothing to fear but fear itself. We have enough to terrify us, and that potential terror is growing. We have to fear the possibility of terrorists who may gain nuclear weapons, and governments like the one in North Korea, ruled by a sociopath, which do not play by the same rules as the rest of the world.

To understand defense in the nuclear age, the student of defense and disarmament is confronted with numerous programs, each technically different, and a sea of acronyms, each subject to a change of name (and acronym) as long as they exist. Garwin was involved in almost all of them. He believes that it is likely impossible to wholly eliminate all nuclear arms from everyone, and probably impossible to develop a completely effective ABM system, but he thinks it is possible and desirable to reduce nuclear weapons to a minimum. MAD would still be in effect because the missiles that would be left would still be devastating.

The US and the old Soviet Union sometimes played nuclear chicken, as in the Cuban Missile Crisis, but both sides eventually backed off for a simple reason: if they did not they would be destroyed, along with their enemy—MAD. Both sides knew that if one attacked the other they too would be at the end of a barrage of nuclear-tipped missiles. This created some interesting situations. If one country, Black, succeeded in building an effective ABM system, or if the other country, White, thought Black did, it would destabilize MAD. If destruction was no longer mutually assured, Black might be willing, depending on the circumstances, to chance a war. If Black developed a new form of intercontinental missile, one that made White's defenses seem unreliable, White might attack first to prevent deployment. In many ways it was a case of perception over reality. If one country *thought* the other was going to destabilize MAD, that would be enough for some action, even if it just meant beefing up an offensive arsenal.

Both sides rely on the other being afraid to start a nuclear war.

For much of the Cold War, the threat of total war came from the air, bombers carrying nuclear and thermonuclear bombs. Then missiles followed, the threat from space. The first reaction in the US as the Cold War began was to rely on the monopoly it had with nuclear weapons. But four years after Hiroshima and Nagasaki, the Soviet Union exploded its first nuclear weapon, a carbon copy of the Nagasaki bomb. The Soviets were no doubt helped by having at least two spies at Los Alamos. With the monopoly gone, the next hope was defense, Garwin said.

What makes the rest of the world susceptible is the development of ICBMs, rockets that can be launched thousands of miles away aimed at a target. Even single missiles could create a problem: a single missile launched from North Korea at Hawaii or the West Coast might actually get through because the interceptor rockets the US defense system would use to try to kill the attacking warhead are not very accurate. That single rocket could be carrying a nuclear warhead.

"We spent many billions of dollars on air defense. We never succeeded to destroy more than fifteen percent of the Soviet bombers in exercises that we conducted ourselves. And finally we fixed on 'deterrence by assured destruction'—that is by having our weapons capable of retaliating even if our country was destroyed," Garwin said.[10] By guaranteeing or at least making the other side think it is a sure thing— that if you strike us we will destroy you—should dissuade any attack.

"There was some air defense of uncertain (but low) effectiveness, but there was no defense against the missile," Garwin said in a speech to the Council on Foreign Relations at a Berlin meeting in 2001. By treaty the Russians were limited to an ABM system that gave some protection to Moscow but could be easily evaded by America overwhelming their system with more missiles. In 1999, the Clinton administration proposed to put twenty ground-based missile interceptors around a Minuteman site in North Dakota or Alaska to defend against a limited attack from North Korea, Iraq, or Iran. Then the number went to a hundred with potential growth to three hundred. The plan eventually was killed as being technically infeasible, Garwin said. Only one of three interceptor tests succeeded, and then only because the target was carrying a large balloon, and it would have been hard for the interceptor to miss it.[11]

Both sides understood this, and in 1972, the US and the Soviet Union signed the treaty banning any country from building nationwide defense systems because it would be destabilizing. It created a fierce domestic battle in America, and probably in the Kremlin as well. The treaty formally endorsed MAD.[12]

In 1972, with the establishment of the Safeguard system, there was the appearance of some protection, albeit of some of the Minuteman missiles and not of the population, but Garwin said there was another multiyear span in which America was again defenseless. "There were no interceptors. There were no radars that could direct interceptors. We were totally exposed to Soviet ICBMs and nuclear weapons."[13]

Opinions on whether an effective ABM system would be dangerously destabilizing or even possible have changed in recent years.

"Certain applications are feasible and stabilizing," Garwin said.

Single-silo defenses against ICBMs are feasible and do not destabilize the balance.

Having ballistic missiles, which behave like cannon balls, and ABMs, which are aimed to destroy, are not uncommon in the world. Some even have guidance. The Israelis deployed US Patriot missiles against missiles fired from Iraq during the 1980 Iraq War, and the Americans and Israelis advertised they were extremely effective, hitting forty-five out of forty-seven missiles, but research by Ted Postol at MIT, using news footage of the interceptors, showed they missed almost every time, making him unpopular in both Washington and Jerusalem. The Israelis used a far more effective missile in the war against rockets from Hamas in 2014. Mostly, Hamas's rockets were not guided. Like Hitler's V-1, they were just pointed in the general direction of Israel and fired. Hezbollah has more sophisticated missiles.

Small rockets do not signal Armageddon. Intercontinental guided missiles do, and against them defense is much less certain.

That's where MAD comes in.

"In a two-superpower world, [MAD] was a pretty good solution," Garwin said. "It has moral and ethical problems. It has problems of stability. It has the difficulty that you might have with an insane leader and a political system incapable of restraining him."

Limiting offensive missiles is complicated. The idea was near "minimum deterrence," agreeing on the minimum number needed to convince an enemy not to attack, and determining how many you need to threaten annihilation, and resisting pressure from the military-industrial complex to build more.

Garwin said the US had had a peculiar problem deciding on targets. All the valuable ones have been targeted by one missile or another, but more weapons keep entering the arsenal. There is nothing left to shoot at.

"I testified to this in 1983 or '84 ... the last nuclear weapon of the 5,000 or so that you have deployed or ready will not cause as much damage as the weapon itself has cost. You have run out of all the important targets."[14] Garwin called for an assessment system so that if one missile fails at launch you retarget another one to get the important target now uncovered, instead of having two or three missiles to an important target, a flexible target policy.

How best to defend the US, Garwin pointed out, was a major political issue. Democrats tended to be blamed for the fact that America was vulnerable, "as if it were their fault."[15] In September 1967, McNamara said that trying to defend against Soviet missiles made no sense. But he announced that the administration of President Johnson would begin the deployment of the Sentinel defense against China's ICBMs. Critics asserted that the advance in science and technology indeed made an effective defense possible against the large number of Soviet ICBM warheads, but that has proven to be untrue, as Garwin has repeatedly said.

Part of the problem in building a defense system is thinking of ways the other side would employ to get around it. A large part of Garwin's work was to think like the other side. Here is what the US is doing to defend against your attack. What do you do to counter its activities?

"We didn't have any missile defense at all."[16]

"ABM is far more than a technical problem [than] the Apollo Program," Garwin wrote, "because the other side, fielding and launching the ICBMs, doesn't want ABM to succeed. Therefore they can make design choices, tactical choices, such as clustering arrival times and preferentially attacking certain targets while leaving others uncovered, and also deploy countermeasures of various sorts to destroy, disable, confuse, overwhelm, under fly, and outthink the defense."[17]

In the half of the *Scientific American* article Garwin wrote, he described the futility of relying on anything except perhaps MAD. It was the kind of thinking Garwin had to do when he came up with dozens of ways the Vietnamese could avoid McNamara's Wall in the Vietnam War, many of which the Viet Cong actually employed. By the time both sides have run out of strategies, the whole idea of an effective defense is rendered useless.

The *Scientific American* article particularly pointed to the Johnson administration's "light" system, called Sentinel, to defend against missiles from China. The government had pretty much lost interest in an umbrella to protect everyone and everything from a Soviet attack by this time. It explored a system aimed at blocking an attack from China or an accidental attack from the Soviet Union, which they assumed would be only a few missiles, not the entire arsenal. The most important protection the US had against China was the certain knowledge in China that the US has "the power not only to destroy completely her entire nuclear offensive forces but to devastate her society as well." Garwin opposed the "light" missile defense, in part because he feared that if the US went through with it, people would get the idea that umbrella protection was possible. All that was needed would be to have the country's scientists get together and invent one, something similar to what Ronald Reagan would say later about the Strategic Defense Initiative—Star Wars—that it was a technologically solvable problem. A complete structure would cost perhaps $40 billion. "Let me make it very clear that the [cost] in itself is not the problem; the penetrability of the proposed shield is the problem," McNamara said about Sentinel.[18] In other words, it can't be done.

The *Scientific American* article showed just how complicated employing an ABM system had to be. Think of the two countries, White and Black, engaged in a nuclear chess game, but a chess game without rules. What if you tried to play chess and you find an opponent's knight, instead of being moved by the prescribed two places, one place, or one place, two places, suddenly moved three in one direction?

In the half of the article he wrote, Garwin named three specific threats to MAD. First, one country or the other actually developed a counterforce system, one that enabled it to incapacitate the other's forces before those missiles got off the ground. Second would be an effective ABM system along with an antiaircraft defensive system. Third, the spread of nuclear and thermonuclear weapons to other countries, something that has already happened since the article was published. To keep aware of what the other country is doing, both the US and the Soviet Union (now Russia) rely on less-than-perfect intelligence. Here is Garwin's war game, the core for his consultation and participation in disarmament and defense theaters, and a demonstration of just how complicated defense is in the age of genocidal weapons.

Black somehow finds a way to negate White's capability to destroy Black by obtaining far more intercontinental missiles, erecting some form of ABM, or developing a way of hitting more than one target with one missile, "multiple independently targeted reentry vehicles (MIRVs)."[19]

White now has to act to maintain MAD. The first thing White would do is to get more missiles and then try to make its own missiles harder to destroy, either by hardening its silos or by putting up its own ABM system. It would also be beneficial for White to set up a system that either launched its missiles just before they are about to be attacked or after an attack that destroyed some number. That, of course, raises the threat of a catastrophic accident; the decision to launch would have to be made in minutes. That strategy still may have some merit, Garwin wrote, because just the idea White has such a procedure might deter Black from an attack, augmenting MAD.

The best way for White to counter the second scenario would be to simply build more missiles to overwhelm Black's ABM system. If White used warheads like MIRVs, Black would have to destroy every warhead to avoid damage. "White could reopen the question of whether he should seek assured destruction solely by means of missiles," Garwin wrote. White might think of more low-altitude bombers or reverting to chemical and biological warfare.

"It does not much matter how assured destruction is achieved. The important thing, as Secretary McNamara has emphasized, is that the other side find it credible," he explained in the article. In his game, if White really thinks Black has developed an effective system, White would have to act accordingly, whether the system is real or not, or whether it is under development or actually deployed.

Garwin also noted that it has been cheaper for White to strengthen its offensive capacity for a massive attack than it is for Black to defend against one. He also pointed out there is a vast difference between a system under development and one that is actually deployed.

How then does one launch a missile attack to get around the defenses? Garwin suggests two strategies. One is to build large missiles capable of carrying many small warheads. All the warheads could be aimed at one target, which would pretty much guarantee that at least one would get through, although each one would require several interceptors to attempt to destroy it. Or, one side could launch a false attack using decoys. The defending nation would fire all its ABMs at the decoys, at which point the attacker would send the real thing.

The day the article came out, Garwin was out of the country. He was somewhat concerned that some government offices were thinking of trying to block publication because, they said, it contained sensitive material.[20] It did not.

"As for Hans Bethe and myself, my own conscience was clear. We revealed no information about the proposed ABM system or the threat that was not generally available. We did bring to this, the expertise for which we were hired by various administrations and which had been honed for many years of working on problems of interest to government and industry," he said.[21]

In April 2005, Garwin addressed the American Philosophical Society in Philadelphia on the growing threat. He called his talk, for which he was introduced by his colleague and sometime critic Freeman Dyson, "Living with Nuclear Weapons: Sixty Years and Counting." He said he "worked a good deal on Mike"; he didn't tell the society he designed it.[22]

Since Oppenheimer lamented being the "destroyer of worlds," the threat has changed, Garwin said.[23] Then the threat was governments obtaining tens of thousands of nuclear weapons. Now it is small groups of terrorists. The governments have far fewer although still too many.

He described Hiroshima as a city "laid level" for miles, the equivalent of a thirty-meter-high tsunami sweeping away everything, and he pointed out that could happen to Philadelphia, New York, Berlin, or Paris. The explosion at Hiroshima was like 10,000 trucks, each loaded with two tons of explosives detonating at the same place and in the same time. A hydrogen bomb could be many times worse.[24]

One of the problems with nuclear weapons is their inefficiency. The most efficient fission bomb—making use of plutonium—is 30 percent efficient, meaning only 30 percent of the plutonium is fissioned. Nagasaki and the one-percent-efficient Hiroshima bombs both weighed four tons and had then to be lugged by B-29s, the largest aircraft the US built at the time. Mike, a fusion bomb, "represented a factor thousand growth in weapon yield over Hiroshima." It turns out, Garwin said, "the most practical use of thermonuclear weapons is to make smaller ones." Mike was a "forty-ton monster," which he simply described as "a fission bomb on one end and a mass of liquid deuterium on the other with experimental apparatus—two-mile-long light pipes for diagnosis."[25] Now nuclear weapons are a few hundred kilograms. Three sit atop each Minuteman III missile, and each can destroy a city. A few kilograms of plutonium could produce a bomb of any energy release, one of the reasons Isidor Rabi and Enrico Fermi called hydrogen bombs "inherently evil."

At the time Oppenheimer died in 1967, the US had 33,000 bombs and the Soviet Union had 40,000. Why?

If the Russians had an air defense system capable of destroying 90 percent of America's attack missiles in the air and many of those yet to be launched, the US would need 10,000 weapons to get beyond Soviet defenses and destroy the Soviet Union, the Pentagon believed. And then there would be the other countries that did not like the US, and, if they had nuclear weapons, they might jump into the fray. You did not need overwhelming force to deter a state from attacking, Garwin told the society. The key word, Garwin emphasized, was "state." To some extent states could be relied on to behave somewhat rationally.

Not all states would be deterred by treaty, Garwin said. North Korea and possibly Iran are two examples. The key to controlling them was to limit access to weapon grade materials, such as the highly enriched uranium and plutonium needed for nuclear and thermonuclear weapons. Plutonium is a human-made element produced by nuclear reactors including power plants. Every reactor produces a quarter ton of it a year, enough for thirty bombs. Every country has the right to nuclear power, and therefore has the capability of pro-ducing materials enough to make bombs. That is where North Korea got its bomb material, from a reactor that produced a small amount of electrical power, but the main purpose of which was to produce plutonium that would readily be separated from the reactor fuel.

The US and Russia have hundreds of tons of the stuff lying around from weapons or from highly enriched uranium from disman-tled arms. Some of the materials in other countries are monitored by the International Atomic Energy Authority, but that organization has only the power to report violations, not to do anything about them.

For over twenty years, the US bought a total of 500 tons of highly enriched uranium from dismantled Soviet and Russian weapons and has been buying the stocks held in Russia, enough for tens of thousands of weapons. It gets blended down in Russia to the point where it is valuable for fuel in power reactors but useless in a nuclear warhead, yet hundreds of tons are still out there. Even small amounts are dangerous.

Garwin said the knowledge barrier to building nuclear weapons is gone. A lot of people know how to build them. And also gone is the political barrier. "A relatively new problem is terrorists and nihilists, a threat that cannot be resolved by deterrence."[26] They are not states. They are freelance.

"Terrorists formerly wanted to only . . . sow terror, but many want to kill people. They don't mind whether they are terrorized," Garwin said, "they just want them dead. And that makes them harder to counter."[27]

In the words of W. S. Gilbert, things are seldom what they seem. According to a CIA report:

"In September 1979 some special security measures were put into effect which indicate that certain elements of the South African Navy were exercising or on alert on 22 September. The harbor and naval base at Simonstown were declared, in a public announcement on 23 August, to be off limits for the period 17–23 September. . . . Also, the

Saldanha naval facility, which included a naval search-and-rescue unit, was suddenly placed on alert for the period 21–23 September."[28]

Early in the morning of September 22, 1979, an American satellite code-named Vela 6911 detected a double flash in the waters off the coast of South Africa. Vela, a DARPA project with Los Alamos participation, was one of a group of satellites specifically designed to detect nuclear tests in the atmosphere, which would have been a violation of the Limited Test Ban Treaty. The flash had all the characteristics of a nuclear explosion, the telltale double flash.

No natural phenomenon could do that, and every time one of the satellites picked up a double flash like that, it turned out to be a nuclear test; except for a large number of "zoo events."

Eventually, the US launched ten Vela satellites to observe the Earth to see the "double-humped flash" indicative of a nuclear explosion. X-rays and gamma rays could also be detected, sure signs of a nuclear explosion. The double flash is caused by first the bright fireball itself, which flares then cools as the volume expands. Then the pressure wave overtakes it, blocking the view by creating brown nitrogen dioxide. The pressure wave moves faster initially and absorbs the light from the fireball. As the pressure wave spreads out, the fireball can be seen again through the clear air for a longer period of time. All big bangs, especially nuclear explosions, are like that. It's also possible to tell how powerful it is, and exactly when it went off.

Vela viewed the entire Earth and thus couldn't say where this blast was—presuming there was a blast. It quickly became a problem.

The flash was detected where the Indian Ocean meets the South Atlantic, halfway to Antarctica, one of the most remote places on the planet. The intelligence community suspected that someone had tested a nuclear weapon in violation of the treaty, and the intelligence community was pretty sure who it was. President Jimmy Carter wrote in his diary, "There was indication of a nuclear explosion in the region of South Africa—either South Africa, Israel using a ship at sea, or nothing."[29] Or, it could have been a collaboration between allies.

What Vela saw is still in dispute. According to Leonard Weiss, writing in the *Bulletin of Atomic Scientists*, most experts think it was an Israeli–South Africa test, but the US government suppressed information, at least in part because it would demonstrate deficiencies in the Limited Test Ban Treaty, and would have an impact on getting ratification of a more comprehensive treaty. Furthermore, putting sanctions on Israel, which would be required if it was found to have violated the treaty, would be a political disaster for any administration. According to Weiss, Richard Nixon had made a deal with the Israelis: the US would not pressure Israel to sign the Nuclear Nonproliferation Treaty if Israel agreed to never admit nuclear weapons or tested one. Weiss explained that the majority of scientists thought it was a bomb test.[30]

Israel had been building a nuclear arsenal since the 1950s with French help. France and Israel were the closest of allies at the time, although relations would later turn chillier when Charles de Gaulle became president and aid stopped. The Israelis, however, had the resources and talent to continue on. Its nuclear facility was and is in Dimona, a small town in the Negev desert south of Beersheba. Because of Israel's small size, there would be no place to safely test a nuclear weapon within its borders.

South Africa, then an international pariah because of its apartheid policies, had been working on nuclear weapons on its own. Israel was the only country willing to sell nuclear-capable missiles to the South Africans and had friendly relations with Pretoria. Both countries claimed they needed such weapons for defense.

Shortly after the incident detected by the satellite, Carter convened a blue-ribbon, eight-person panel to review what was known about the event. The panel, headed by Garwin's colleague Jack Ruina, an MIT engineer and former head of DARPA, included Nobel laureate Luis Alvarez; Garwin's friend and ally in the disarmament choir Stanford physicist Wolfgang "Pief" Panofsky; and Garwin. It was to report to Frank Press, Carter's science advisor.

The incident was kept secret until October 25, 1979, when ABC reporter John Scally reported it after a Pentagon leak. Carter was constantly briefed on the panel's investigation, all of which was held behind closed doors. The findings, released in a 1980 report, were—and to some extent—still are classified. The report declared it was likely *not* a nuclear event but did not rule it out.

According to Weiss, that conclusion was not universally accepted. Scientists at national laboratories like Los Alamos and Sandia produced evidence to the contrary, and Donald Kerr, head of the Nuclear Intelligence Panel, simply said, "We had no doubt it was a bomb."[31] Scientists at Los Alamos wrote: "We have constructed what we believe to be a plausible model for a low-yield nuclear explosion that could have produced the observed Alert 747 signal [the 1979 Vela signal]." The National Security Council first said it was a bomb, then said it was "agnostic."

Several laboratories disagreed.

The White House released just the conclusion of the Ruina report, but several days later, according to Weiss, the National Research Laboratory produced a 300-page report concluding it was a bomb. Weiss said that when Ronald Reagan took over the White House, Weiss, then working for Senator John Glenn's Senate subcommittee on nonproliferation, was ordered not to contradict the Ruina report in public testimony.[32]

Part of the problem in determining the cause of the flashes was an inconsistency in the two light measuring devices called bhangmeters on the satellite, which may have malfunctioned. It showed up in a "timing tick," a difference in detection time of the two onboard bhangmeters that gave a false reading. One of the instruments was beginning to show its age, engineers said, which may have accounted for the tick. Weiss points out, however, that the satellite, whatever its age, had detected every known nuclear test since its launch in 1969.[33]

If it wasn't a test, what was it? The Ruina panel came up with two possibilities, including a glint of light from debris floating in the ocean or the collision of the satellite with a micrometeorite.

Although he claimed politics has played a role in the suppression of evidence, Weiss wrote that there was no evidence that anyone on the panel, including Garwin, succumbed to political pressure or "did anything other than give their genuine scientific opinion."[34] The problem was, he said, the panel was restricted to a scientific analysis and was not given whatever the intelligence agencies had. Others were not so kind. One former member of the Nuclear Regulatory Commission, Victor Galinsky, charged they were politically motivated.

Investigative reporter Seymour Hersch, who authored a book on the Israeli nuclear program and reported on it for the *New Yorker*, suggested that Israel would have had to test at least three bombs to be sure they worked unless someone gave them design codes, which also would be a violation of the Nuclear Nonproliferation Treaty. There is no evidence that anyone did. Hersch wrote he was told the test off Africa was a low-yield nuclear artillery shell. If it was a nuclear weapon, it produced two to three kilotons, a relatively tiny device. It also could have been the nuclear trigger for a thermonuclear device, like the primary in the Teller-Ulam design.

Garwin has been the leading defender of the report based on the bhangmeter anomaly. He points out there has been no corroborative evidence to support that a nuclear weapon test took place. A few days after the signal, the air force sent planes over the area looking for radioactivity but found none. In general, you do not shoot off an atomic bomb without leaving evidence behind.

Alvarez later wrote that the usual ways of reading the satellite records didn't agree as well as usual:

Drawing on my bubble chamber experience, I asked to see a selection of the satellites' "zoo-ons," events so strange they belonged in a zoo. ... Since their records were stored on computer tape they needed only a week to put their zoo together, Rich [Richard Muller], Dick [Garwin] and I found a steady degradation in record quality among those zoo-ons from confirmed explosions to events at which no one would look twice. Although the event we were studying had some of the characteristics of a nuclear explosion, only one of the two satellite sensors recorded it. Moreover, there was no indication from earlier or later records that the sensor that failed to record the event was malfunctioning. Both sensors looked at a large area of the Earth's surface so it was hard to believe that one sensor could see a nuclear blast and the other could not.[35]

On the other hand, the astronomical center at Arecibo, Puerto Rico, recorded an unusual ionospheric wave at about the same time as the flashes were detected. Others, looking at the same data, said the second sensor did in fact record something. Hersch wrote it was an Israeli test. Another theory, that the device was South African, that the Israelis taught South Africa how to build a device in exchange for uranium, has since been discounted.

At the CIA, days after the flash, Garwin, Harold Agnew (recent director of Los Alamos), and Steve Lukasik (former DARPA director) were shown what data had been collected.[36] The three concluded they couldn't tell with the information they had. But they had to tell Carter something.

"I said, 'Well, you can't really tell,'" Garwin said. "Given this evidence and the fact there was no other evidence at the moment—50-50. No other evidence came in, so that makes it much less likely." What made it complicated, he said, was that it happened in the one place on Earth other satellites would not have picked up a test to corroborate what this satellite recorded.[37]

Most scientists appear to agree it was an Israel–South Africa operation as circumstantial evidence began to appear.

"In my opinion, it never occurred," Garwin said. "The only evidence for it is that we didn't see anything except for an anomalous signal from one of the satellites."[38]

CHAPTER THIRTEEN

TREATY

For more than sixty years, the United States has been negotiating to reduce the number of nuclear weapons in the world, mostly with Russia (nee Soviet Union), and more recently with other nuclear powers, and especially to reduce the probability of their use. The negotiations have largely succeeded, and humanity has not been wiped out. Nuclear weapons still exist in quantities capable of destroying human civilization. Nonetheless, the international community has seen progress. The main treaties include:

Partial Nuclear Test Ban Treaty—signed in 1963. The treaty banned testing of nuclear weapons in the atmosphere, outer space, and in the ocean, but did not prohibit testing underground. The original parties were the United Kingdom, the United States, and the Soviet Union. The ultimate goal of the treaty was to eliminate all testing of nuclear weapons entirely as quickly as possible. The impetus for the treaty was radioactive fallout from aboveground and underwater testing, which presented serious health risks to civilians. Additionally, in the air and water, any contamination could spread and harm the citizens of other countries. Underground testing was not included because of the difficulty of verification, mostly how to differentiate a nuclear test with a seismic event.

Treaty on the Nonproliferation of Nuclear Weapons—signed in 1970.

There were two parts to this treaty. For countries that already had nuclear weapons, they vowed not to sell, share, or encourage non-nuclear countries to manufacture or obtain nuclear weapons. For the non-nuclear weaponized countries, they agreed not to accept or pursue nuclear weapons. The treaty also established "nothing in this Treaty shall be interpreted as affecting the inalienable right of all the Parties to the Treaty to develop research, production and use of nuclear energy for peaceful purposes without discrimination"[1] as long as they did not violate the sharing of weapons and weapon information. The 190 signatures of the treaty agreed not to acquire nuclear weapons and agreed to promote nuclear energy. The ultimate purpose is the reduction of nuclear weapons worldwide. Only four countries have refused to sign the treaty: India, Israel, Pakistan, and South Sudan. There were 191 signatures, but North Korea withdrew in 2003. India and Pakistan have nuclear weapons; Israel probably has nuclear weapons and possibly thermonuclear devices.

Anti-Ballistic Missile Treaty—signed in 1972 by the United States and the Soviet Union. The treaty limited ballistic missile defense systems. Each country was limited to two defense centers, one around its capital, the other at least 1,300 km away, protecting an ICBM launch area. That was eventually reduced to only one area for each country. The US chose a Safeguard site in North Dakota to defend Minuteman silos, while the Soviets chose to defend Moscow. By limiting the defendable areas allowed for each country, the treaty institutionalized MAD and came at a time when technological advances in missile warheads, like MIRVs, made a secure system infeasible. SDI (Star Wars) would have breached the agreement, but after first vehemently opposing Star Wars, Moscow dropped its opposition, concluding, like had most American scientists, that it would never work. After thirty years, the US withdrew from the ABM Treaty in 2002.

Strategic Arms Limitation Treaty (SALT I–II)—SALT I was signed in 1972 at the same time as the Anti-Ballistic Missile Treaty, and negotiations for SALT II began soon thereafter.

SALT I froze the number of strategic missile launchers at the existing levels, and put limits on submarine-based missiles. The range of ICBMs also was limited by the treaty, as was the number of protected sites. SALT I expired after five years. The SALT II talks went on for seven years before a final agreement was made. SALT II limited the manufacture of new weapons and accounted for necessary specifications based on how the United States and Soviet Union chose to develop their nuclear programs. SALT II was signed in 1979 but was never ratified by the US Senate after the Cuban Missile Crisis and the Soviet invasion of Afghanistan.

Strategic Arms Reduction Treaty (START)—the START program was really a series of four separate treaties that overlapped, were ineffective, or abandoned due to political tensions on both sides. START I, a 1991 agreement between the two superpowers, reduced the number of strategic nuclear weapons by 80 percent, the largest arms reduction treaty in history. Each country was limited to 6,000 warheads on 1,600 missiles. After the long negotiations for START I, the Soviet Union dissolved less than a year after the treaty was signed, calling into question its validity. The new states became part of the treaty; however the process delayed the treaty going into effect by more than three years. Kazakhstan, Belarus, and Ukraine disposed of their nuclear arsenals by returning the warheads to Russia and joined the nonproliferation treaty as non-nuclear states. Possibly the most significant advancement in START was the introduction of invasive verification methods: both parties were required to allow for on-site inspections and extensive data sharing.

START II talks began shortly after START I went into effect, and was signed in 1993, and later ratified by the United States and Russia. However START II never went into effect because Russia made its ratification contingent on the Anti-Ballistic Missile Treaty staying in effect and the US withdrew from ABM in 2002. The Treaty

Between the United States of America and the Russian Federation on Strategic Offensive Reductions (SORT) replaced START II with weaker regulations and was in effect from 2003–2011. SORT was superseded by the current New START treaty, which took effect in 2011, and is expected to remain effective until 2021 at the earliest.

Garwin was involved in every treaty, almost always as a consultant. He never actually negotiated a treaty. Often he just made his opinions known to friends in the government who, it would seem, valued his advice. Frequently he would write position papers. Other times he consulted as part of a JASON project or as a member of PSAC.[2] The JASONs were, in fact, more involved than most people realized. Sometimes it was indirectly, as with various satellites that constituted the "national technical means of verification" enshrined and protected by the SALT agreements. The success of these satellites and the Vela system through the years encouraged the US to sign the Limited Test Ban Treaty; the ambiguity over the Atlantic test was an anomaly. JASON studied the proposals for the government, which would become the Comprehensive Test Ban Treaty, and they gave support and formalized the idea that substantial explosions, of the kind that meant a militarily significant test, could be detected, and the ones that could not be detected were not devices that posed a military threat, Garwin said. "The fact we could not ourselves do low-yield explosions would not impair national security." Neither would the limitation affect US offensive capabilities.

Garwin served on PSAC during the Partial Test Ban Treaty. The eventual treaty was derived from the Conference of Experts in Geneva where Garwin *kibitzed* while he was working on the surprise attack negotiations in the summer of 1958. The actual treaty negotiations were in Moscow, and information on what was happening there was transmitted to the negotiation support team in Washington, that included the White House National Security Council and, in those days, PSAC. Garwin was on vacation in Paris but was asked urgently to return to Washington to help review the treaty to see if the treaty emerging "would suit the President's purposes."[3]

There is a protocol for starting negotiations for a treaty, although the protocol is frequently skipped over. The idea behind a treaty almost always starts at the top, Garwin said, with a decision from the president that such a treaty would be in the national interest. The process then moves to the US Department of State and the foreign ministry of the other country or countries involved. (Bilateral treaties, of course, means two countries negotiating; multilateral treaties, which are more complicated, means more than one country). The process can become tortuous because each party involved has its own value system and will only sign a treaty if it gets what it wants, or a reasonable facsimile, Garwin said. Bilateral treaties allow more horse trading.

Negotiations start out not with the other country but domestically, with whomever the stakeholders are. Their world is going to change if the treaty is signed. Whatever they've done before the treaty or what they have to do after it will change their jobs. "Whatever they do is going to go away as a result of the treaty."[4]

In every negotiation, each country is trying to get the most from other countries while minimizing the limitations placed on itself. For example, post-industrial countries are more likely to advocate a strict fossil-fuel regulation because their economy has moved past burning fossil fuels and such a restriction would not significantly affect the country while improving the environment for everyone. Most treaties include restriction elements and action elements. For example, an environmental treaty could have countries burn less fossil fuels while also forcing them to improve their pollution cleanup practices.

Once it is established that both countries are interested in forming a bilateral treaty, each side forms a working group to meet in a neutral location like Geneva or, alternatively, in one of the home countries. In the US, support groups are formed in Washington, and, if the other party is Russia, they have a support team in Moscow. The support group talks to domestic constituencies like the military and intelligence community to decide what the boundaries are for the negotiators. When the negotiators hammer out the language, the support groups at home can provide input on what provision can or cannot be done.

While this setup allows for both sides to get the information they need, it is also a successful negotiation technique. If both parties at the table have to answer to an outside authority, it is harder for anyone to take advantage of the group. It also sets the groups up to be willing to work together to achieve something acceptable instead of fighting it out in the room. This method can also decrease the likelihood of personal politics to become involved.

The ironic method both sides use is to each bring an outrageous extreme to the negotiation table so that the parties can come to some "option C" in the middle. Both sides are pushed to go for the most outrageous positions because if only one party uses an extreme view, then the final decision will be skewed in that direction. However if both views are too far away, then negotiations can fail because the parties present themselves as fundamentally opposed regardless of the similarities of the actual positions.

"Each side has its own position it can back down to," Garwin said.[5] Sometimes the US negotiators were sent out without orders, which happened in several treaty negotiations including the surprise attack meeting in Geneva, and the Threshold Test Ban Treaty of 1974, which was called to limit the energy release permitted in an underground nuclear test explosion. The US negotiators were sent out without anyone telling them what the limitations would be. Ultimately they settled on forbidding testing bombs with yields above 150 kilotons. PSAC and Garwin were very active in the ABM treaty and the SALT I agreement, pointing out the problems of ballistic missile defense and how it could be overcome by the Soviets.

"You wouldn't be protected, but you would have to say you were being protected to get the money for it," Garwin said. "You would be worse off. . . . You would have to build up the destructive capabilities of the world, you would have to put it on a hair-trigger alert. It would be very bad. Finally we persuaded the US and many in the military—the contractors weren't as easy to persuade. They are in business to make money. . . . That's not evil although sometimes it has a very bad influence."[6]

The US wanted to limit offense as well as defense; the Russians wanted to stick with limiting defense but were ultimately induced to change their position and limit their offense as well.

The US and the Soviet Union gathered at a summit meeting at Glassboro, New Jersey, in June 1967, while the Six Day War raged between Israel and most of the rest of the Middle East. The Soviet position at the summit was that "defense is only good," Garwin said, and that anything a nation needs to defend itself is a good thing.[7] At that point, scientists involved in the Pugwash Movement got involved to encourage more talks. Pugwash is an international organization of scientists formed from a manifesto by Bertrand Russell, the pacifist mathematician, and Albert Einstein, the pacifist physicist, who pointed out that the presence of nuclear weapons, particularly the hydrogen bomb Garwin helped design, really changed world dynamics and attention must be paid. What hadn't changed, Einstein and Russell wrote, is the way humans think.

Pugwash provided the cover for American and Russian scientists to meet. The Ford Foundation also got involved as did the American Academy of Arts and Sciences. Jack Ruina was involved, and so was Garwin. The appropriate Russians also were in on it, so the groundwork was there.

However, back channels can short circuit the whole thing. That happens.

The Russian scientists maintained the party line, "how can you be against defense?" But at a Pugwash meeting, a Russian scientist, using a mixed metaphor, said missile defense was like Moses striking the ground with his staff to get water and "all the devils of the underworld merged from the wounds of the earth." He had been converted, a hint that the Soviets could be persuaded. Nixon, a Republican, had more freedom of action than a Democrat would, so he and Kissinger went to Moscow, cutting the State Department negotiators out of the action by sending the US ambassador to the Soviet Union out of Moscow. Kissinger could be in charge, and he and Soviet Ambassador to the US Anatoly Dobrynin had established a back channel to communicate, sometimes using the phone, or sometimes the ambassador came to visit Kissinger's office. They decided what the treaty would be about.[8]

Sid Drell, Garwin, and others had met monthly with Kissinger in the Situation Room of the White House as informal advisors preparing top-secret "think pieces," on such things as limiting bombers. That's how people like the national security advisor or secretary of state sometimes get their information and recommendations, papers from experts. In many cases, the consultants give a list of options. In one instance, Kissinger asked the group to investigate the multiple warheads of multiple independently targetable reentry vehicles (MIRVs). They presented five options, which Kissinger then ignored. In his memoirs, Garwin said, Kissinger admitted he should have paid more attention to their advice. He never mentioned the group of advisors.[9]

Kissinger's first response when he got a report on missile defense from one of the group, however, was "we must get PSAC out of strategy." Garwin said they were not in fact doing strategy at all; they were simply assessing the technology of missile defense.[10]

"That's a good example of how I was heavily involved in the ABM treaty," Garwin said.

But on one occasion, frustrated by bureaucracy, Garwin took matters in his own hands and formulated what could have become a treaty on his own.

The dissolution of the Soviet Union came as a surprise to practically everyone, including, perhaps, the people of the Soviet Union. The Western intelligence community was clueless. When the Berlin Wall fell, Georgy Arbatov, a Russian political scientist, told Garwin, "We have taken away your enemy."[11] That left a serious threat, however. The Soviet Union had tens of thousands of nuclear weapons all over Eastern Europe in places like the Ukraine, Belarus, and Kazakhstan. Besides loaded on bombers, there were ICBMs with nuclear warheads in silos. Submarines also were armed with nukes. The great fear was that terrorists, or a rogue country, would take control of the weapons or that someone would steal the nuclear fuel and thereby arm terrorists. And there were many thousands of relatively portable nuclear weapons to be delivered by relatively small aircraft.

Meanwhile, the Soviet Union was falling apart. Garwin said when he traveled there he took along a suitcase full of sausages to give to friends, not to mention hundreds of dollar bills in US currency. He had also arranged for many Russian scientists to work for the US while still in Russia on projects aimed mostly at keeping them busy and out of trouble. Garwin got an agreement at IBM that researchers at the Research Division could use as much as 10 percent of their research budget on Russian scientists.

"You could buy them for 10 percent of the price of a scientist in the United States," Garwin said.

The weapons were a real danger to world security. They now belonged to Russia. Wolfgang "Pief" Panofsky headed the Committee on International Security and Arms Control (CISAC) at the National Academy of Sciences, which met twice a year with their contemporaries in Russia. They went to work on the problem as well. It was concluded that there were many things that could be done with the weapons' nuclear fuel, but most would be far more costly than was feasible and none of them satisfactory.

CISAC at the National Academy of Sciences came up with two rules for handling the fuel. One, a nuclear weapons standard to make plutonium safe from theft by keeping it as protected as the bombs they came from. It's just as dangerous. Two, spent fuel standards after the plutonium is processed to make them less dangerous so that it would be no more desirable a target for theft than fuel from a nuclear facility.

"I felt it would take a long time for the [CISAC] study to be complete and I wanted to work with the world community on this," Garwin said. "I took the initiative on my own to explore with the NATO scientific element whether they would fund a NATO scientific workshop on the management and disposition of excess weapon plutonium."

Garwin and two other scientists went to Paris in 1994 to talk to people in the French military and to legislators trying to convince the French to stop nuclear testing. In 1992, the French said their ability to simulate nuclear weapons with a computer meant they could add weapons without testing. They were very proud of that, Garwin said. It turned out they were wrong. They did more calculations and found they needed a series of tests at their Pacific site. They continued to carry out atmospheric testing until 1996. That was internationally unpopular to the point that the Australians banned French wine. Garwin opposed the testing but thought it might have a silver lining: the French could perhaps be induced to joining a comprehensive treaty banning all testing.[12]

Garwin decided to call his own international nongovernmental meeting of nuclear states to discuss a solution to the plutonium problem. He did not get any support from the US government but didn't get any opposition either. The French did not respond fast enough, but a British think tank agreed to sponsor the study.

NATO agreed to pick up the expenses. Russians came, so did the English, and "some crazy people from Los Alamos," and they all gave technical papers.[13] The Russians stayed over for two days because they apparently took the problem seriously. The conference report came out before the National Academy report.

The ideas from the conference were forwarded to people in both Moscow and Washington. When the CISAC report came out, there was a joint US-Russian commission, and treaties were based on the work. The Russians wanted to take the excess plutonium for their breeder reactors that generated electric power. No US power companies wanted to use fuel from weapons in America because of public relations impact, and also because it would be a daunting cost. The battle continues in Washington to this day.

The Russians brought the warheads back to Russian storage facilities. The rockets the warheads sat on are now being used to launch US satellites. Thirty-four tons of plutonium on each side was agreed to be surplus and slated for disposal. Garwin said a once highly classified fact was that the average amount of plutonium in an American nuclear weapon averaged four kilograms (8.8 pounds), about the size of a baseball, and you can make 250 weapons per ton, about 8,500 nuclear weapons' worth, "not small change."[14]

No results came of Garwin's freelancing. "Ran afoul of the desire on both sides to protect their freedom of action, and the great secrecy surrounding the 'overhead reconnaissance' satellites," he wrote.[15]

CHAPTER FOURTEEN

STAR WARS

The origin of the Strategic Defense Initiative (SDI), better known as Star Wars, is a more than a bit confusing. Some said that President Reagan was at a lecture at Lawrence Livermore National Laboratory when Edward Teller gave a speech on a complicated, high-tech missile defense system. Or was it a 1979 Reagan visit to the North American Aerospace Defense Command (NORAD) in Colorado, a huge facility carved into a mountain that monitored the skies and space looking for an attack? The writer Francis Fitzgerald is probably correct that it might have been a science fiction film Ronald Reagan starred in during his career, *Murder in the Air*, a 1940 movie in which an American agent is supposed to protect a secret weapon that is designed to destroy enemy planes in the air. Then again, Fitzgerald points out, there was Alfred Hitchcock's 1966 *Torn Curtain* about the development of an antimissile weapon in which Paul Newman declares, "We will produce a defensive weapon that will make all nuclear weapons obsolete and thereby abolish the terror of nuclear warfare."[1] Reagan often confused reality with the movies.

What is clear was that he disliked the whole concept of MAD, which he described as a mutual suicide pact. He also used the metaphor of two men pointing pistols at the other's head. Ted Postol, who was a science advisor to the chief of naval operations, said that MAD was never really a policy; it was an "existential condition." It simply was how things were.[2]

On March 23, 1983, Reagan astonished Washington and the science establishment with a public announcement that he was initiating a long-range research and development program to build a defensive weapon system that would make all nuclear weapons obsolete and thereby abolish the terror of nuclear warfare. He slipped it into the last few paragraphs of his speech after reminding everyone of the dire threat America faced from the Soviet Union. The media immediately dubbed it "Star Wars" after the George Lucas film.

The Joint Chiefs of Staff apparently had been informed of the president's plan, but no one else in the Pentagon was, nor were the secretaries of defense and state until just before the speech. The latter were important because MAD had been the condition that kept the United States and the Soviet Union from destroying each other, and SDI, if it worked, was going to destabilize that. Everyone, including Reagan and Teller, understood SDI would require enormous leaps in technology to be successful. Nonetheless, elements in the Pentagon were jubilant. Postol walked into a Pentagon office the next morning and saw some officers dancing around their desks. One of them shouted, "We are going to defend lives, not avenge lives!"

Postol asked how he was going to do that. The technology wasn't there, and it was not in the immediate future. He said he got no answer.

Later he was asked to brief the chief, Admiral James Watkins. He gave Watkins several reasons why it was a bad idea, pointing out the damage one Soviet submarine parked off the West Coast could do to a defense system. About two-thirds through the briefing, Watkins got up and walked out of the room. Postol never briefed him again.

Postol later found out that Watkins was one of the people who had gone to see Reagan to vouch for the potential of such a plan. He was motivated by a deep religious concern for the possibility of a war that would kill millions.

"How the hell do you make a momentous decision like that without checking with your science advisor?" Postol said.

Just before Reagan's speech, Ed Frieman, a Princeton physicist who was vice chairman of the White House Science Council, was called to Washington from vacation. The president's staff was cabling an advance copy of the speech to cabinet members and to Margaret Thatcher in London. Frieman and others who worked on the panel told Reagan candidly that they did not know how to do what he wanted. That made no difference to Reagan.

"I need to protect my country," the president said.[3]

Reagan called on the scientists who had produced the atomic and hydrogen bombs to now work on a way of rendering them impotent. According to Fitzgerald, scientists and defense experts who were invited to a White House dinner that night were mostly appalled. Not only would it be decades before the technology he was envisioning would be ready—if ever—but it would cost a fortune and probably trigger a new arms race. If one side destabilizes MAD, the other side has to react, and whatever can be said about MAD, it worked.

Shortly after the broadcast, the administration set up a committee headed by Garwin's colleague James Fletcher, and coheaded by his close friend Harold Agnew to report on the feasibility of SDI. The committee published a set of unclassified volumes that broke the various assets of SDI into individual problems and concluded that none were impossible. Supporters ran with those judgments, supporting the plan. At the same time, a young staff member, Simon F. Worden, commissioned supporting arguments from Greg Canavan at Los Alamos, Lowell Wood (a Teller protégé), and Robert Jastrow of Dartmouth.

In 1984, Reagan demonstrated he was serious. He established the Strategic Defense Initiative Organization (SDIO) and asked Congress for $26 billion for five years of funding.[4] To wide astonishment, he offered to share the technology developed for SDI with the Soviet Union and American allies. That later became changed to share *the benefits* of the technology, an offer Garwin suggested was something like sharing the benefits of slavery. "The slave owners benefited; the slaves benefitted," he said.[5] The Soviets were unimpressed. As Fitzgerald points out, SDI may have been Reagan's "greatest triumph as an actor-storyteller."[6] The American people believed Star Wars could be done.

Simply, the plan was to neutralize all Soviet missiles by attacking them at various parts of their flight. There would be laser battle stations in space and on the ground. The beams of pure, intense light would hit the missiles as they came. As a backup, defensive American missiles would be launched to attack the Soviet missiles. The silos for those missiles would be around major targets, including the missile silos the US would use to counterattack. Defense missiles would be protecting the offensive missiles. Infrared detectors would provide the trigger.[7] The system relied on weapons not yet invented, including nuclear X-ray lasers, subatomic particle beams, computer-guided projectiles fired by rail guns, and space drones, all of it run by a supercomputer, which also didn't exist. Mostly the weapons were aimed at destroying the incoming missiles high in space to minimize damage on the ground. Some were land based, some in space, and some would be placed on submarines. One other problem: it would require so much energy that dozens of nuclear power stations would have to be built to run it.

Like using a magnifying glass to start a fire, the particle beam weapon would focus atomic or subatomic particles into an intense beam that could burn a hole in the fuel tank of a missile. It's a good theory; in particle accelerators like the one at CERN, they are able to get particles up to the required speed to make a particle beam weapon, but there are several important differences between the lab setting and what is necessary for a weapon to function in an uncontrolled environment.

Discussing the details for a particle beam always raises more questions than it answers. If you have a beam on the ground, the amount of energy required to burn a hole through the atmosphere and hit the target without the particles getting deflected is immense. A possible solution is to use a less powerful beam, but that requires a longer exposure time and tracking capabilities to keep the beam focused on the target as it moves through the sky. However this type of tracking and focusing system is currently outside of our capabilities. There also would need to be measures to detect nonthreatening objects, to know if the object has been destroyed, and account for defensive measures like a magnet that could deflect the beam.

To avoid the problems of making a beam strong enough to work in the atmosphere, some people suggested that the system could be in space orbiting Earth. The first problem with this is that it would be difficult to have the particle beam in the right place at the right time to defend specific locations. Then there is the problem of the Earth's magnetic field, which deflects charged particles. There would also need to be communication between the ground and the satellite, which could be jammed and leave the particle beam useless.

On top of every technological problem, there is no evidence that the beam weapon would be the most effective. It would be much easier to intercept a Soviet rocket by instigating a collision with another object.

X-ray lasers were the brainchild of Teller and Woods in the 1980s. A young physicist at Lawrence Livermore, Peter Hagelstein had actually invented one for his dissertation. It used X-rays instead of visible light. Essentially, the laser was designed to be used in research, opening opportunities to study such phenomenon as plasma. Eventually the lasers permitted insights into such things as the molecular scale structure of materials—natural and synthesized—a useful research tool. Even biologists have found ways to use the lasers by looking at things like the structure of cells.[8]

A rail gun consists of two parallel metal rails and an electric power supply. When a projectile is placed between the rails, it completes the circuit so that the electricity is flowing from the rail through the projectile and back through the rails. The current causes a large force to accelerate the projectile along the rails as the current continues to flow through it. Because the propellant is electric instead of a chemical reaction or explosion, it is possible to increase the speed rapidly with no added weight from fuel. The US Navy has recorded projectiles traveling as fast as Mach 6.[9] Theoretically one aims the rails at the target.

The idea was to aim these powerful X-ray laser beams at incoming missiles as soon as the system spotted them. They would require a thermonuclear explosion for a source of the light. Such a laser was actually tested during a weapons test in 1979, in a project called Project Excalibur. Teller once bragged a setup about the size of a desk could wipe out all of the Soviet Union's launched missiles.

Garwin said the creators "scarfed" up every possible technology and threw them into the project.[10]

SDI would radically alter America's whole defense structure, and Reagan asked Ed Frieman's group to see if any of the newer technologies would have an impact on the feasibility of the system. Frieman was a Democrat in the White House. "I felt like a spy," he said.[11]

The council took classified materials and what was known by the public, studied it all, and concluded none of the known technology would make SDI successful. The clear implication was that SDI still was not achievable.

"Teller at that point went into orbit and demanded my report be suppressed," Frieman said. The technology Teller believed should make SDI work was, the panel concluded, useless. The panel was wrong, Teller said. "I mean it just created holy hell."[12] Frieman relied on David Packard, former deputy secretary of defense under Nixon, to intercede with Reagan to calm down relations with the panel, and relied on Garwin to reassess the report and see if he thought it was valid.

"Well, during those times, I talked to Dick [Garwin]," Frieman said. "I didn't of course tell him about what was going on with Edward. That was internal White House stuff. But I tried to get from him as much as I could, and his assessment of the technology to see whether the conclusion that we were stating that there was no way that any of those weapons could influence strategic modernization over these years, whether that was in fact a correct statement or not. So we had these discussions. I relied on Dick. And the discussions weren't long, where he'd sit down and lecture me. It was, I'd see him in the hall and I'd ask a question, or we'd discuss it over dinner or whatever. But I can only tell you from my perspective, having Dick's imprimatur on . . . these weapons . . . was incredibly important, because I trusted his judgment on these scientific and technological issues really far above almost anybody else."[13]

Garwin corroborated the findings—the technology wouldn't change the circumstances.

"So following the speech, there was this huge group of people brought together under DOD auspices, one group after another to study it," Frieman said. "And Dick was all over the place in those things, and just tore every argument to shreds. It finally got to the point where Dick and a few people were really arguing . . . with some of the guys at Los Alamos, particularly Greg Canavan." Canavan and Garwin would joust for years, sometimes in the pages of *Physics Today* and *Nature*.

During the debate, Garwin and Hans Bethe, now also enlisted in the anti-SDI force, blasted the whole idea of Star Wars. Defense Secretary Caspar Weinberger gave a paper Garwin and Bethe coauthored to Canavan. Canavan's specialty, Canavan said, was not to discredit reports like this but to find mistakes in the algebra. "I'd correct their algebra and they'd go away," he said. It was one way of finding out if they were "serious." The serious ones always came back.[14]

"It was a typical Garwin-Bethe report," Canavan said. "Well done. They knew what the important assumptions were, knew what mattered, and knew what didn't matter. I went through the body of the report. They laid out here are the problems you had to face. 'I can't tear this down,'" Canavan said to himself. "'I'm screwed.' . . . Here's physics. Here's math. Therefore it won't work."

It had appendices, said Canavan. They were pretty good too, he found. Then he found Appendix C.

Appendix C was about how many lasers would be needed to do "boost stage defense." Hitting enemy rockets while they were still rising was the best defense because all the action was far from the homeland. It was all about lasers.

"Five years before I concluded that lasers were not a player in space because they are big, clumsy, they can't move," Canavan said, "and they are very vulnerable. A nuclear warhead could pop them all off at a greater distance than they can shoot."

What the defense system *was* going to be was a secret, Canavan said, so neither Garwin nor Bethe knew what was planned and just assumed there'd be lasers, and then they wrote why they wouldn't work. For convenience they supposed all the missile launches from the Soviet Union would be from a single point, which made the calculations easier. Canavan's study had the missiles spread over a large area. He said the algebra was wrong. Garwin and Bethe believed it would take 2,500 lasers to do the job if they were not destroyed. The right answer was twenty-five, according to Canavan.

Unlike others whose work was pushed off by Canavan because Canavan found problems with their algebra, Garwin and Bethe did not go away. They came back with a new, lower number, 1,250. Canavan knew it's still twenty-five. Another time they came in with six hundred. Someone drew a chart showing Garwin and Bethe's numbers going down while Canavan's stayed the same. Then in one more meeting, someone cracked, and the Garwin-Bethe estimate would go below zero.

Canavan said Garwin and Bethe finally "went away." When the report for the Union of Concerned Scientists was published, Garwin and Bethe's Appendix C had been mysteriously eliminated. Garwin said, however, that he had corrected Canavan's algebra in a *Nature* article.[15]

Why didn't Canavan just tell the two that lasers had already been eliminated? His report was a secret, and one of the rules in this world of secrets is that if someone makes a mistake describing the secret in public, you can't correct them or otherwise the person would know there is a correct answer, and even the existence of a correct answer is a secret.[16]

"I would say that Dick's role in trying to bring some scientific credibility to showing that this was not a way to solve the world's problems was crucial," Frieman said. "He wrote about it. . . . He was on one panel after another. And in fact, he gave the administration fits because he was deadly opposed to this; thought it was wrong. . . . It didn't bring it to an end, clearly, because these current guys are building a system we don't need, and it won't work, but other than that, it's all fine."[17]

Later, Dan Ford helped arrange a television commercial, "Star Wars," based on a statement from Garwin and Bethe opposing Star Wars, which was sponsored by the Union of Concerned Scientists. The UCS only had the money to put it on air once and only in Washington. However, one showing was all that was required because many news organizations picked it up and began showing it nationwide free of charge.

"That helped a lot, too," Frieman said. "Before that commercial, it was always the Reagan Star Wars program. After that commercial, it was always referred to as 'the controversial Reagan Star Wars.' . . . It just took one word to change the tenor of the debate."[18]

Probably to no one's surprise, Frieman was fired.

The debate became heated and personal. In an incident at a scientific meeting in Italy, Garwin's patriotism was questioned.

"I think it was 1983, I guess," Garwin said. "And in fact both Flora Lewis, who used to be a foreign correspondent for the *New York Times*, and Mary McGrory of the *Washington Post* were there and so was not only Edward Teller and some other people from Livermore, but Dixie Lee Ray who had been head of the Atomic Energy Commission or whatever it was called. I got into arguments. I explained why Star Wars wouldn't work and it would be against our interests. There were Russians there. Dixie Lee Ray accused me of being a traitor because I opposed something that the government was wanting to do. Edward Teller told my wife he didn't 'know what had happened to Dick. He's become a war monger because he's not interested in defense, only in offense.'"[19]

At one time, Garwin said, Teller proposed small nuclear anti-tank weapons that would kill the soldiers inside by radiation. Garwin asked why that was necessary since there already were high-explosive weapons that could take out a tank and the inhabitants inside. Teller responded that he saw no point in rejecting a possible weapon without first trying to see if it was feasible.

"Edward was a very smart person," Garwin said, "with lots of empathy but when he was against you on some program he would attack you personally. That was a character flaw in my opinion."

Despite being on opposite sides of many defense-related issues, and despite the almost universal shunning of Teller by the physics community, Garwin and Teller remained friends. Several people commented that the affection and respect between them was obvious when the two were seen together. Garwin had no hesitancy in admitting that was so. But largely that was when they were not fighting. Then Teller turned vicious.

The battle surrounding Star Wars waged both in the halls of Congress and in the media. Three Democratic senators reported in 1988 that SDI would cost $170 billion just to start, would not be operational until the turn of the twenty-first century, and even if it did get up and running it was unlikely to kill more than 16 percent of incoming Soviet missiles. A second phase would cost at least $541 billion, for a total cost of three-quarters of a *trillion* 1980 dollars. The report was put together by aides to Senators J. Bennett Johnston of Louisiana, Dale Bumpers of Arkansas, and William Proxmire of Wisconsin. The report came after the Pentagon's Defense Science Board had just suggested breaking SDI up into its components and emphasizing ground-based interceptors. Another study suggested that the Soviet Union could find easy ways around SDI, and still another, this by the Office of Technology Assessment, predicted a "catastrophic failure" the one and only time it had to be used.[20]

Resistance to SDI came from everywhere, even the Pentagon.

Garwin reported that US Air Force Director of Plans Major General John A. Shaud observed that the system wouldn't work unless there was an air defense system to go along with it, adding hugely to the cost. "'If you're going to fix the roof, you don't want to leave the doors and windows open,'" Garwin quoted Shaud as saying.[21]

In the course of the battle, Garwin often debated Star Wars in public forums, often with SDI's head, US Air Force General James Abrahamson. Sometimes his ally was the astronomer Carl Sagan. In one session, in November 18, 1987, Garwin, Abrahamson, and Sagan were joined by former Assistant Secretary of Defense Richard Perle and moderated by Massachusetts's congressman Edward Markey.

Perle began pointing out that the US had absolutely no defense against a massive attack, which was true, even from a solitary missile, which was partly true, a situation "unprecedented in history."[22] He pointed out that SDI was an R&D program, one designed to develop the technology for the future, and there were no plans to deploy the system. He said he was surprised that men of science would reject such a plan since it was essentially just research. The US could not and should not try to stop the march of technology, and as to MAD, he said he saw no difference between active defense (SDI) and passive defense, such as bomb shelters.

Garwin explained that the rationale for SDI kept changing, from one that would protect the entire continent to one that would protect military targets only. Civilian targets would go unprotected, because, backers said, there would be no military advantage to hitting cities. SDI would enhance deterrence, backers claimed. Garwin didn't use the term, but he was describing "mission drift," changing the purpose of the project to keep the project goals relevant or obtainable. Neither goal was obtainable.

Abrahamson said critics of the program also changed the debate by redefining SDI for their own purposes. They kept raising wrong questions and supplying the wrong answers, essentially describing a different project, a straw man argument. Meanwhile, the Soviet Union was aggressively exploring the same technologies, Abrahamson said.[23]

Sagan, as was his wont, went for the dramatic. There were 60,000 nuclear weapons, including 25,000 strategic weapons, at the time, enough for two countries to destroy each other's homelands, a "grotesque disproportionality," he said. There were, he pointed out, only 2,300 cities in the world. By best estimates, SDI would be able to stop, at most, 90 percent of the missiles fired at the US, meaning about 1,000 would get through. "The shield is leaky," he said.[24] A hundred million people could die around the world from such an attack.

What often isn't clear is that Garwin did not oppose spending money for research that might lead to Star Wars, agreeing somewhat with Perle's position. He was opposed to spending money to actually implement the program because he was convinced it wouldn't work. In an interview on CBS's *60 Minutes*, in a quote that was edited out, he explained in reference to charged particle beam weapons that the money ($24 million) "is money well spent, even if it shows us that there is no promise there because so often you spend money not to build something but to find out that it can't be done, that the other side can't do it either. So I support the program."[25]

Garwin once wrote:

> People who live in the glass house of deterrence should not throw stones. Furthermore, they should not develop better ways of throwing stones, even if they never plan to throw those stones themselves. Yet a commitment to the development of defenses (even if we later decide not to deploy them) will cause us great difficulty even if we ascribe to the Soviet Union the capability to accomplish those things that we have demonstrated or simply promised. Because technology is useful in *defeating* strategic defenses, an analysis of SDI demonstrates once again that optimism about the future of technology in no way ensures that the technology, if realized, will make us more secure.[26]

Historians debate whether Star Wars brought about the fall of the Soviet Union when it became clear to Moscow it had insufficient resources to compete. Frieman and Garwin had a mutual Russian friend, Roald Sagdeev, one of the Soviet experts trying to counter SDI. "After it was all over," Frieman said, "I talked to Roald, because this debate was going on. I said, 'You know, Dick and I and others were trying to stop all of this.' And I said, 'How did you guys react? Is there some truth to this statement that SDI had this effect?' And he said, 'I've got to tell you . . .' He said, 'You guys oversold it, but we overbought it.'"[27]

CHAPTER FIFTEEN

GRAVITY

Physicist have spent much of the last century trying to prove Albert Einstein was right in his revolutionary theories—or better yet, that he was wrong. So, when someone publishes a paper experimentally proving a major finding of the theories of relativity, the world pays attention.

In 1969, when Joseph Weber, a Talmudic scholar, Annapolis graduate, World War II officer, and professor of physics at the University of Maryland, said he had good evidence gravity waves exist, attention was paid.

Physicists were dubious, not that gravity waves exist—although there were some who doubted that—but that Weber had found them. Among the skeptics was Richard Garwin.

Weber set up an experiment using two six-ton bars of aluminum with sensitive detectors attached, one at College Park on the Maryland campus, the other at the Argonne National Laboratory in Illinois. The bars were suspended in a vacuum by a steel cable. If the bars detected gravity waves, they would chime two octaves above middle C, something like an electric xylophone, Garwin said.[1] Weber claimed he was registering ten a day. Nobel prizes were given for less.

He would eventually build a different version that went to the moon, the Lunar Surface Gravimeter. His massive detectors are known as Weber bars.

To Einstein, gravity waves were ripples in space-time, like ripples in a pond. It would take something cataclysmic to produce any such detectable waves, which scientists believe correspond to a relative shift of one one-thousandth the diameter of a proton over a distance of a kilometer or more. To many, Weber's experiment was too insensitive for the task.

Attempts to replicate the experiment elsewhere were failing. Curious, Garwin went to College Park to look around. What he saw did not impress him, a "waste of time and money," he said later.[2] Not only was the experiment set up wrong, he believed, but Weber had blown the math. He was looking for a non-random event that would hit both detectors simultaneously but did not properly account statistically for apparently simultaneous random events.

"I've never known Dick [Garwin] to make a quantitative error, in so far as you estimate something," said Walter Munk, "and sometimes I miss it by a factor of 10 or 100 or so. He doesn't do that. He has such a good feeling for things."[3]

Back at IBM, Garwin and James Levine decided to see if they could do it better. "We scrounged up parts, aluminum bars from the stockroom, and built what came to be known as a Weber bar, but with provable sensitivity better than that of Joe Weber." The conclusion was simple: "Weber could not have detected gravity waves," Garwin concluded.[4]

He described Weber's experiment as an "astonishing comedy of errors." Researchers in Germany and Italy, using even better equipment, were also coming up blank. At MIT, Rainer Weiss, also trying to duplicate Weber, also was dubious about Weber's work.

"He had made a huge and powerful claim. And I began to realize maybe this was wrong, and maybe even his idea of how it works was wrong."[5] Weiss began building new experiments using interferometers, devices that use beams of light—in this case, lasers—to make minute measurements.

Garwin and Levine published their findings in *Physical Review Letters*, the same prestigious journal that first published Weber, essentially destroying Weber's conclusion.[6]

By this time, Weber had doubled down on his assertion, and enlisted a collaborator, whom Garwin thought a very good physicist, David Douglass of the University of Rochester in New York, who set up another bar.

Garwin and Levine worked up another paper listing the errors in Weber's work, including the fact that Weber's report of his work with

Douglass had included a major clock error. One recorded the time of gravity wave events on US Eastern Time, the other on Greenwich Mean Time without noticing, Garwin said.[7]

"Levine and I worked on gravity waves until we had confidently disproven the tremendous 'discovery' that there was so much power in gravity waves. If Weber was right, the universe would have been converted entirely to gravity-wave energy within 100 million years, but our sensitivity was far from being able to detect any of the potential real sources of [gravity waves]—for instance the in-spiral of a pair of solar-mass neutron stars or black holes." In other words, if gravity waves were so strong Weber could see them with his equipment and the universe would have disappeared billions of years ago.[8]

At a meeting at Cambridge, Douglass and Weber sat in the front row when Garwin was to make his oral presentation. The two men had signed a contract saying that neither would say anything without the other's permission. Garwin said he didn't want to give his presentation and embarrass the two men in public. If one agreed to stand up and admit he was in error, Garwin would sit down.

Weber wouldn't, and Douglass perhaps couldn't. Garwin gave the talk, and the Weber gravity detector experiment went into the scientific trash heap.

Not everyone cheered Garwin. Freeman Dyson thought he went too far.

"Weber was an imaginative scientist who did the wrong experiment," Dyson said. "For that Garwin has no mercy. Something wrong is wrong and he would just grind you to the ground for it and he did with Joe Weber. I was sad about that."[9] If Weber and Douglass proved anything, however, it was that maybe it was possible to detect gravity waves. Perhaps the technology wasn't yet advanced enough when Weber tried it.

According to Garwin, "Weber is just such a character that he has not said, 'No, I never did see a gravity wave.' And the National Science Foundation, unfortunately, which funded that work, is not man enough to clean the record, which they should."[10]

Weber lost his funding, although he continued his research on his own.

Back at MIT, Weiss and his coworkers were continuing to experiment with detecting waves with interferometers, with machines getting bigger and bigger. Eventually, the California Institute of Technology and physicist Kip Thorne got involved.

Then the money ran out. It was during the Vietnam War, and gravitation and cosmology were not high on the priority list. The physicists stayed with the work and came up with a proposal to the National Science Foundation for a new type of interferometer gravity wave detector, but once more, money was scarce. The work was at a standstill. Enter Richard Garwin again. According to Weiss, Garwin interceded with NSF, telling them, "If you're going to persist with this, you'd better have a real study." NSF agreed. It was tough love.[11]

Out of those efforts came the Laser Interferometer Gravitational Wave Observatory, otherwise known as LIGO, actually two detectors, one in Hanford, Washington, the other in Livingston, Louisiana. The detectors make use of the purity of laser light. Anything, no matter how minuscule, that interfered with that light would be detected.

The LIGO structures consist of two two-and-a-half-mile-long structures set perpendicular to each other. Lasers are shot down the length of each structure.

LIGO takes advantage of the phenomenon found at the event horizon around black holes. As you approach a black hole, you would get stretched and elongated. Your feet and your head would grow farther apart the closer you get, and you would become thinner. If a gravity wave passes through LIGO, one detector would get longer, the other shorter by an almost imperceptible amount. The longer the detector, the greater the difference in light-path length along the two arms, the reason LIGO has such long arms.[12]

On September 14, 2015, LIGO scientists reported that LIGO had detected unmistakable gravity waves. It was a momentous observation, proving Einstein right again. The source of the waves was stupendous: 1.3 billion years ago, two large black holes collided, forming one large black hole, seventy times more massive than our sun. They were part of a binary black hole system pulled toward each other at almost the speed of light. In their final collision, they shot out the gravity wave detected in Washington and Louisiana.

Three months later, the LIGO researchers detected a second wave. "Although the signal is weaker than the first one," they announced in a press release, "this new finding has also been confirmed with a confidence level of over 99.99999 percent."[13]

Weber died in 2000 still maintaining he had detected gravity waves.

The science fiction writer Arthur C. Clarke is universally credited as the inventor of the communications satellite, a machine in synchronous orbit that hovers over the same place above the surface of the Earth and relays messages, phone calls, television programs, etc. Actually, he was beaten to it by almost a century by the American writer Edward Everett Hale, and a story called "Brick Moon," published in the *Atlantic* in 1870. Hale essentially described a space station. The idea that something could be kept in one orbit by gravity, of course, goes back to Isaac Newton who knew that the moon stayed up for the same reason apples fall.

Shortly after World War II, the US Air Force set up the RAND Corporation to do research and development, and one of its first reports was to call for using satellites for reconnaissance in support of military activity. The Cold War was coming. When Russians had launched Sputnik, Garwin said, most people had no idea what to make of it.[14] President Eisenhower had to give press conferences to assure everyone Sputnik wouldn't fall and kill someone. Sputnik itself wasn't a threat, but missiles appeared capable of dropping atomic bombs on American cities. In 1958, PSAC created a two-person subcommittee, Edwin Land, later founder of Polaroid, and Edward Purcell of Harvard, to assess how the United States should respond to Sputnik. The report helped energize US satellite efforts.

Clearly watching the Soviet Union from the sky was the answer. But how?

One idea the Land panel came up with was floating balloons in the atmosphere's "sound duct" to detect missile firings. The sound duct is that layer of the atmosphere where sound tends to collect because of a decrease in air temperature. They would essentially listen to the roar of launched missiles. That never happened.[15] There were more realistic ideas.

They decided they needed to know more about what was happening in the Soviet Union, but at the time, satellite technology wasn't advanced enough to do the job.

Land and Purcell, and a couple of others sponsored by the Science Advisory Committee, encouraged the CIA to create the U-2 spy plane, put together by Lockheed's Skunk Works in California. They had it flying in eighteen months, Garwin said.[16] Because the U-2 could only reach 65,000 feet, it was still in the range of Soviet surface-to-air missiles, so the Skunk Works created the A-12, which could fly at Mach 3 at 90,000 feet, well out of the capabilities of the Soviet defense system. The CIA thinks it may be the source for many UFO "sightings" that were made in the 1950s. Plus its engines did weird things and made pulses of multicolored lights. Most importantly, it flew faster and higher than any plane should, and no one could acknowledge it existed, so if even an air traffic controller were to ask, "What the hell is that?" he'd get no answer. "We have no idea what you are talking about. Perhaps your instruments are faulty."

When Francis Gary Powers was shot down in 1960 in his U-2, the United States said it would no longer overfly the Soviet Union—at least with American pilots. The plane was turned over to pilots from allies, perhaps Taiwanese, to overlay China, Garwin said.[17]

"The Land panel also realized that the SR-71's days would be numbered," Garwin said, "and we ought to have a satellite observation program. The Air Force had at RAND a program that was essentially a TV tube that could be orbited, and sent down it signals by radio the way weather satellites do." Resolution wasn't very good because of their altitude, the amount they have to cover, and the purpose of the satellite. They used batteries for power.

The US had a satellite in development called Vanguard, which was supposed to be the first space satellite ever launched, but it kept spectacularly failing—on television—and Sputnik got there first. The project had been deliberately put under civilian control because the Soviets were surely going to protest if an American military satellite regularly sailed over its territory.

Nonetheless, the US appeared to be in no rush.

The navy, air force, and army all proposed satellites of their own. The army proposal had the backing of Wernher von Braun, but the navy won the competition and the Naval Research Laboratory had control of the project.

On March 17, 1958, a Vanguard satellite was finally put in orbit. It is still up there while all the other early satellites—Soviet and American—have long been gone.

James Baker of Harvard, who was designer of most of the optics for the planes, designed lenses for what became the Corona satellites in 1960. The first had thirteen successive failures before it got into orbit and operate, but Eisenhower kept loyal to the program until the engineers finally got it right. One hundred forty-five "buckets" containing 2.1 million feet of exposed film were returned to the Earth by an inventive process in which the buckets dropped from the satellites on parachutes and were caught in midair by a trapeze hanging out the back of cargo aircraft over the ocean—a prodigious piece of flying if nothing else. The film then was sent to Eastman Kodak in Rochester, New York, for processing, duplication, and distribution to the people who wanted to see the pictures. The images could tell how tall buildings were, could spot tanks and chimneys, Garwin said.

On August 18, 1960, a Corona spacecraft took the first pictures of the Soviet Union from space. Garwin had a hand in all the surveillance satellites, and his work is almost entirely classified. He doesn't talk about it.

The US never admitted it was using satellites for intelligence until 1978, and Corona itself was top secret until 1995, long after it went out of service. From an intelligence standpoint, it was a bonanza, peeking into "denied areas" in China and Russia. According to a once-secret report, one mission "provided more photographic coverage of the Soviet Union than all previous U-2 missions."[18] Cameras got better, and resolution improved.

"Severodvinsk, the main Soviet construction site for ballistic-missile-carrying submarines was first seen by CORONA," the CIA said. "Now it was possible to monitor the launching of each new class of submarines and follow it through deployment to operational bases. Similarly, one could observe Soviet construction and deployment of the ocean-going surface fleet. Coverage of aircraft factories and airbases provided an inventory of bomber and fighter forces. Great strides were also made in compiling an improved Soviet ground order of battle."[19]

The early Corona satellites only had one bucket. The images went horizon to horizon, strips 700 miles long, 200 miles wide. They were run by battery-operated mechanical timers, like the timers one plugs into outlets to turn lights on and off.

By 1972, Corona was replaced by two other film satellite systems, Gambit and Hexagon. Hexagon had ten buckets and could stay up for months, sending a bucket every week, unless the intelligence community thought something was going on they had to see, and then they could order up a drop. Committees of users would decide when to drop a bucket of images.

Even the name Corona had been classified. Garwin was one of the relatively few authorized to know about it. In 1995—after it was decided to declassify Corona—Garwin spoke, representing the White House at a two-day public meeting at George Washington University. The project contractors and the CIA also made presentations. The meeting was emotional, Garwin said.[20]

Those working on the project could not tell their families what they were working on. Those in "black" intelligence programs—totally hidden and off the budget—couldn't even say where they really worked, not even tell their spouses or parents. They would get up in the morning, dress, get in the car, and drive away, or wait on the sidewalk until a nondescript car pulled up and took them away. Their families had no idea where they went, what they did, when they would come back, or even *if* they would come back. Emergency telephone calls went through an operator. It could be stressful.

"My family knew I was involved in Washington and did some secret work," Garwin said. "I never talked about it."

There was some unhappiness because the project workers thought they would go to their graves without anyone knowing what they had done, and they were very proud of their work. Now, for the first time, the Corona people could brag and tell their families what they had been working on for the past few years: they were protecting their country by spying on the Soviets during the height of the Cold War from space.

"It was very emotional to hear people talking freely about what they had done," Garwin remembered.

Newer systems, all digital with solar panels for electricity, now monitor Russia. Much of the satellite arsenal is part of the "black budget," so secret it appears nowhere in print in the public record. Even the names of the satellites are secret, many operated by the CIA. Garwin worked on most of them and has won an award from the CIA for his work. Some scan the ground for images; some look for radar signals. He does not discuss his contribution. He was named one of the Ten Founders of National Reconnaissance by the National Reconnaissance Office.

CHAPTER SIXTEEN

HEALTH, PANDEMICS

In 1918, an influenza epidemic killed between twenty and forty million people in what was the worst pandemic in recorded human history. More people died in one year than in the four years of the Black Death in the fourteenth century. Most epidemiologists think it is likely to happen again. So does Richard Garwin, who has made studying pandemics something of a sidelight. He served on the Permanent Monitoring Panel on Terrorism of the World Federation of Scientists, chairing a subgroup on mitigation in 2006, and gave a presentation on pandemics at a meeting in Erice, Italy. Since biological warfare is most likely to erupt in an act of terrorism, his interest is not unwarranted. What can be learned from pandemics like flu and smallpox can help protect the public from a biological attack.

Studying the possibility of a deliberate pandemic rose partly because of the flu outbreaks in the 2010s. Most experts think the most likely cause would be from the influenza virus H5N1, although a recurrence of H1N1, the so-called Spanish flu of 1918, still is a possibility. It is not probable that a terrorist would use influenza as a weapon. That disease would spread around the world too quickly, and would eventually strike societies the terrorists would "esteem." Anthrax and tularemia are more likely military weapons since they are not very contagious and don't spread readily from person to person. Garwin said even with advances in medicine, such an outbreak of smallpox would be a disaster, largely because of globalization with people moving around the world with ease.[1]

Garwin and his committee said there were three ways to prevent or combat such an epidemic: personal protective measures, vaccine to build immunity, and pharmaceuticals to mitigate symptoms and kill the bacteria or microbes doing the damage. Personal measures would include hand washing, face masks, and improved air filtration. Getting people to follow instructions would be difficult, and while some of the measures make sense, Garwin said there is no strong scientific evidence they work. Attacks on cities could be devastating, particularly to people living in the release point who would have no time to do anything.[2]

In 2006, Garwin was a coauthor of a paper in *Science*, along with scientists from the National Center for Disaster Preparedness and the Alfred P. Sloan Foundation, and explained that personal efforts were still the best defense. It would take four to six months to get enough vaccine out, and antivirals would only be a backup. The problem is what is not known.

"Recent attempts to identify the most effective nonpharmacological interventions have revealed that these measures have a thin science base. For example," they wrote, "it is uncertain whether influenza transmission from person to person is primarily by large droplets or by fine particles." That matters because the answer would determine just how far you should stand from someone else to avoid infection. It would also determine what kind of face masks would be required. Scientists aren't even sure why it is so hard to break the chain of infections of influenza, considering that the flu is not exceedingly infectious.[3]

Washing hands seems to make sense, but there is no science to support that it would work with influenza. "There are no readily available compendia of best practices or even comprehensive databases of community epidemiological data which might help to design the most effective interventions." More studies are needed, and seasonal influenza, the illness that strikes relatively harmlessly every year, would be a good base for research, the three authors said in the *Science* paper. Improved air-handling systems, room-sized fans, portable air-filtration units, and perhaps even physical barriers might be tested. New buildings should have protections built in; old buildings should be retrofitted.

Garwin pointed out that the fairly new custom of coughing into the inside of your elbow rather than in your hands might help prevent the spread of the virus, "as long as you don't drape your arm over someone's shoulder."[4]

In the summer of 2011, Johns Hopkins University Medical Institutions, one of the world's premier medical centers, announced a revolutionary advance in medical care: the Epic Project. For a sum estimated to be almost $1 billion, Hopkins was going to finally put the medical records of its patients, the data from its laboratories, medical tests, images, as well as its business and finance onto digital media where it would be available to whoever needs it across Hopkins's huge institution. Hopkins is spread across Maryland. A patient seeing a doctor at one of its satellite buildings in Lutherville can go to the main medical campus in East Baltimore to see a specialist who can ask for an MRI in Hunt Valley, and all the doctors involved in the care can call up the test results and the images, as well as the patient's personal information, everything needed to give the best care possible. So can the patient. The network links six academic and community hospitals, thirty-nine outpatient centers, data on the 2.8 million patients who visit each year, the 115,000 hospital admissions, and 40,000 employees. It even does the billing.

In 1968, Garwin proposed just such a system.

He could not know how computer technology would evolve in the half century since he designed his system. He could not know about the rise of desktop computers, tablets, the cloud, cheap memory, Bluetooth, or the Internet. But what he described in general terms in his paper is almost exactly what the Epic Corporation installed at Hopkins almost fifty years later. Medicine is only one discipline transformed by computers, and while it finally happened, it wasn't easy. Hopkins wasn't breaking new ground; other medical centers had already installed a similar system. But the technology was moving at its own pace, and the medical profession can be quite conservative.

In the 1960s, Garwin was a member of the new technologies panel of the National Health Manpower Commission organized in

1966 by Lyndon Johnson to look at new technologies for the nation's healthcare system. The panel was chaired by Albert Wheelon, who had been the first deputy director of science and technology for the CIA. He and Garwin both were on the delegation to the surprise attack meeting, so they knew each other. In the early 1950s, Garwin also had done some work on health issues at IBM, which of course wanted to sell computers to hospitals. The commission visited hospitals, local healthcare centers, and Kaiser Permanente facilities in Los Angeles. They went to England and Sweden to study how computer technology was used there. At the Karolinska hospital in Sweden, they saw new patients given wristbands with bar coding on them so that when blood samples were taken, patients could be positively identified. That was considered the state of the art in electronic medical art then and is still widely employed today.

But even today, there are doctor offices in which entire walls are taken up with folders containing patient records, many of them color coded for identification, and when a patient signs in, someone has to find the folder and ready it for the doctor. The information is generally useless outside that office unless it is faxed over or otherwise transmitted individually.

"It was clear what was needed," Garwin said, "so I put my thoughts together to make them more widely known." The result was a paper published in *Public Health Reports* in 1968 called "Impact of Information-Handling Systems on Quality and Access to Health Care."[5]

Garwin wrote that the issue had two aspects: the healthcare system needed a way of keeping the records of individual patients, and it needed a general reference service, sort of a Wikipedia instantly available to physicians.

The limitations were extreme. Computer memory, which now is so inexpensive it is hardly noted, was extremely expensive then, making it practically impossible to store images. Communications between computers usually involved telephone modems, with data reduced to audible chirping. A good X-ray image would have 100 million bits without compression. The limit was 300 bits per second, if the line was clean. The best Garwin could say about the modems was that they made lovely doorstops.[6] At corporations and institutions like hospitals, individual monitors plugged into mainframe computers (mostly made by IBM), not the most efficient way of doing things. Personal computers were just on the horizon.

The key to his system was an automated laboratory hooked to the network. If a lab could produce a blood sample analysis in minutes instead of half a day, it would save everyone money, including reducing hospital stays. A patient has a blood sample taken, and then the lab analyzes it and posts the results almost immediately into the computer network. The information is constantly accessible to every resource the patient needs for the center, the physician, the hospital, the counselor, etc. The data bank would contain all the patient's data, past and present. Garwin knew the technology wasn't up to this yet, mostly because of the cost of communications then. Eventually, he predicted, the files would be accessible from anywhere in the world. (He believed that someday patients would carry all that information in a "microphotographic record." Now it might be described as a chip.)[7]

The accumulated data, he wrote, would also be useful for medical research, a huge bank of information not possible otherwise, including "statistics of disease, variability of certain measures, and indices to health." The data would also have "safeguards to health."[8]

"Inexpensive, routine automated testing and the existence of a patient data bank may be able to provide baseline information on each individual for comparison with subsequent screening or for diagnoses of disease," Garwin wrote. It could spot patterns not easily identified without the database—Big Data—although he did not call it that. Researchers in Britain are currently using just such a database in their efforts to understand Alzheimer's disease, using tens of thousands of patient records from their National Health Service.

He proposed a medical referral service, providing "up-to-date material readily accessible to the physician through a convenient communications terminal. It might contain, among other things, the United States Pharmacopeia, synonyms, drug prices, specialists and their hours of practice and a medical reference library." It might offer "machine-aided diagnosis" and perhaps the ability to send electrocardiographic signals and images to distant eyes and ears. All that information would follow a patient no matter where he or she went.

"Beyond the bare bones of the system . . . the possibilities are fascinating. The medical reference service can provide the physician with such diverse services as a medial newsletter tailored to his own interests or refresher courses in his specialty."

We call that the Internet.

The system would be tested rigorously often to prevent degradation, and probably would need to be replaced every five years. He did, after all, work for IBM. The system would have to be changed as technology advanced.

"I believe that our nation is often faced with the choice between very large investments in the current way of doing business and substantially smaller investments in more efficient and economical ways of providing better service," he wrote.

CHAPTER SEVENTEEN

FAR OUT

Few events in American history can match the assassination of President John F. Kennedy for birthing conspiracy theories. One of the most prominent is the acoustical theory, which stated that there was a second gunman firing at Kennedy on the grassy knoll by Dealey Plaza. A policeman's microphone, probably on a motorcycle, was stuck "on" for long minutes before and after the shots were fired. Acoustic experts analyzing the Dallas Police Department's recordings interpreted sounds on the recordings as gunshots and their echoes from buildings, and then deduced the location of the microphone as it moved through Dealey Plaza. They concluded that the muzzle blasts originated from the grassy knoll and that there had to be a second sniper. The FBI didn't believe it. They were convinced Lee Harvey Oswald acted alone and there was no second gunman. The theory achieved some gravitas when the United States House Select Committee on Assassinations issued a report claiming there was more than a 95 percent chance that a second sniper was involved.[1] The committee's conclusion was based on six sound impulses on the recordings and on radio calls between dispatchers and the officers. That report immediately caused the predictable uproar.

The National Academy of Sciences was asked to put together a study to go over the evidence and see if the congressional committee was right. Twelve scientists were chosen, including two Nobel laureates, Luis Alvarez of Berkeley and Norman Ramsey of Harvard—and Richard Garwin. Garwin had worked previously with Ramsey.

"That's how these things work," Garwin said. "We did the usual things. We tried to understand what was going on. We looked at the data. People had ideas, especially Ramsey and Alvarez."[2]

A rock musician in Ohio wrote the committee and said he could hear cross talk between the two recordings and hear instructions to the police after the shooting. The recordings could not have captured shots because overlapping the sounds people thought suspicious were commands to the police to head to the hospital.

Alvarez had the idea of looking at the FBI voiceprints and could identify features in them so he could determine the timing between the two channels. "I was pretty happy with this but I decided computers would do a better job," Garwin said, and that would call for the Fast Fourier Transform to get the noise out of the signals. The voiceprint is a picture of the Fourier Transform.

One of the recordings was not at a constant speed, so matching the two proved difficult. Garwin took the prints up to IBM and called in speech-processing scientists. Together they were able to match the recordings, especially the sounds on both of them.

There was no second shooter.

A whole cottage industry then sprang up to criticize the NAS report. In 2001, an agricultural statistician, B. D. Thomas, published a report attacking the conclusion. Some of the members of the committee were uninterested in responding, but Ramsey "was a real bear" in urging a response.

Garwin went back to work. By then, he said, there were massive improvements in computers, with PCs running faster than the mainframe used previously. Garwin reprogrammed the analysis for the PC. An IBM physicist and friend, Ralph Linsker, collaborated. They obtained what Garwin called "beautiful and new results."

"We were quite confident the original result was correct," Garwin said. The sounds of the shots were not on the recordings.

Putting conspiracy theories to death is a game of whack-a-mole. Thomas repeated his charges in a book but has since been mostly ignored except by conspiracy fans, who think the Ramsey report was fraudulent.

"It is possible for a technological group like ours, of mavericks and scientists, whose reputations depended on honesty, to be mistaken, yes, but cheating, no," Garwin said.

Lee Harvey Oswald was the lone assassin.[3]

In the beginning of the Cold War, when technology amplified paranoia, both the Soviets and the West were afraid of aircraft flying near or at their territory. The Soviet Union was complaining that Strategic Air Command bombers were flying in the polar regions and appeared threatening, and SAC was monitoring every flight of every Russian bomber they could see. Were the planes part of an attack? What were they doing? Where were they going? The Soviets were particularly concerned when the United States sent B-47s, six-engine high-flying bombers, to probe Soviet defenses. America was just as excited about the Tupolev bombers that flew toward Canada and the United States over the North Pole. And then there were submarines. The mutual fear was tangible.

In the summer of 1958, Garwin, happily ensconced at IBM in New York City, had an idea: what if all aircraft and ships carried beacons—transponders—that automatically transmitted identification and location so each could be tracked? He made his proposal in the publication of the *Bulletin of the Atomic Scientists*. The polymath in him was expressing itself.

Garwin suggested that the United Nations create an agency that would continuously monitor and interrogate all ships and planes (and spacecraft) that operate in international airspace or waters from ground stations around the world. The agency would sell simple transponders, and any vessels traveling in international space would be required to have one. The vessels would be tracked via satellite relay from a dozen sites around the world; none of them needed to be in Russia or the US, Garwin wrote.[4] Computers would interrogate the beacons, identify the vessels, track them, and give out the information to any country that asked. They could turn off the beacons when traveling in their home countries. Any vessel traveling without a beacon was fair game in international space and could be destroyed or impounded.

The plan would have the additional benefit of providing the vessels navigational information and would aid in air traffic control. Positions would be determined by triangulation by two or more ground stations. The beacons would automatically locate crashes, a capability the black boxes aboard airliners now have, or ships in distress.

The ships would file information of where they were going and how; aircraft would file flight plans.

Every ship would be interrogated once an hour, every plane in five-minute intervals. Interrogation would be in random order to eliminate fake responses. Safeguards would have to be installed to prevent accidents or to keep someone from gaming the system.

"At present we might observe a flight of Soviet planes leaving Russia but we have no way to identify a hostile intent until they are over Canada or the US," Garwin wrote. "Add to this the real uncertainty in their detection at long range and the fact that we must not attack these planes until they reach Canada or Alaska, and we see that our only response must be a general SAC and fighter alert (and vice versa with the Russians)."[5]

If the planes carried beacons, the US would know how many there were, where they were, and where they were headed, greatly reducing the chances of an international incident. If the planes did not carry the beacons, or if they did not have registered flight plans, SAC could attack.

Garwin thought his plan would please the Soviets because they would allay fears of the SAC flights. And, of course, he worked out the science. One of the advantages of his plan—this was in 1958— was that it did not require the use of great amounts of data. Given 30,000 planes and 500,000 ships, the ground stations would transmit a maximum of fifty bits per plane times 100 planes per second. The plane would send back a few bits in a tenth of a second. Communications satellites would handle the network and provide the time of receipt of the signals—a kind of inverse GPS providing the system precision location of the vehicles.

"This is a peacetime program and is recognized as such by all participants," Garwin wrote. "To jam the satellite is forbidden by international agreement, and the sudden jamming of the transmission costs us no information that we have at present and is a sure sign of bad intentions on the part of someone, since it is a serious interference with the work of an important international agency."[6]

The defense air tracking system was not Garwin's only plunge into the astounding complexity of air traffic control. In March 1971, Garwin, as a chair of the Ad Hoc Air Traffic Control Panel of PSAC, reported to Richard Nixon on ways to improve America's air traffic control system, which PSAC said was suffering from years of neglect. Commercial airliners were stacked up over the nation's airports. It wasn't the technology, the report said, and it wasn't that the US didn't have airports that were "among the finest in the world."[7] It was that the number of aircraft was overwhelming the system. Delays were rampant, and there was no flexibility. If a storm struck Chicago, flights all over the country were affected, and recovery to normality was difficult. In 1970, it cost the airline industry $400 million, turning a year that would have been profitable into a loss year.

The group to study the problem was formed in 1970 by PSAC, with Garwin head of the aircraft panel, then the Air Traffic Control Panel. In 1971, PSAC issued a report strongly recommending an all-satellite system for navigation, communication, and surveillance. PSAC said aircraft could be equipped with transceivers for navigation for $900 that would be accurate to within 100 feet if the computation was done on the ground. Airliners would do the computation onboard for devices that cost about $2,100 per plane. Rather than criticize the analysis of the report, the Department of Transportation hired the Institute for Defense Analysis to do a competing study, and it came up with a price of $200,000 per airliner for the cost of the onboard equipment.

"This is not a prescription for progress," Garwin wrote. He complained that there was no technical content in the IDA report—just an estimate of how much it would cost for old technology to provide the location accuracy and timeliness that the new technology could do for $1,000–2,000. Think GPS.[8]

As is fitting for a science committee, they recommended more funding and research.

They recommended data links between the ground and the air instead of a voice link with an onboard printer to receive messages. They also recommended a satellite-based surveillance and navigation system, something similar to what Garwin had suggested to watch the Russians. The system also should be automated.

With air travel becoming ever more congested, sooner or later someone had to think of a better way of handling airports. Since many of the larger cities are on the coast and most of the rest of the land is developed with the things that make them a city, some cities, such as Hong Kong and Kyoto, have built out on a landfill, "which immediately begins to sink," Garwin said.[9] In the US, San Francisco and Boston also are on a landfill.

He came up with a revolutionary idea in 1969—a floating airport, nicknamed FLAIR. No one has built one yet—it may be just ahead of its time. Acting with an engineering company, Weidlinger, Consulting Engineers in New York, he and Paul Weidlinger designed one that would sit twenty miles offshore, have runways long enough for conventional aircraft to land and take off, and wouldn't sink if a fully loaded Airbus 380 or Boeing 747 landed on it. Modern jumbo jets can weigh a million pounds.

"Land a plane on a platform, it's going to tip. It may not sink, but it will tip," Garwin said.[10] Garwin's design originated in his PSAC Aircraft Panel study of transportation in the Northeast Corridor under a contract from the Department of Transportation on a $25,000 contract.

One idea was to make use of vertical landing and takeoff aircraft, Garwin said. If a plane is landing horizontally as normal planes do, it has to come in on a specific flight path, with congestion, approaching a long runway. Takeoffs would be just as regimented. Vertical-operating aircraft could come in at any angle, as long as traffic control kept them from crashing into each other. But the technology of huge vertical takeoff and landing aircraft was—and still is—far in the future. The innovation was to be automatic landing in less than thirty seconds, thus saving the fuel that would otherwise have been burned in a standard five-minute landing requirement.

Putting the airport out to sea would mean less congestion, and technically, it could be moved even farther out if the city expanded with landfill. Noise problems would be eliminated since the airport would be away from the people. And current aircraft could use it. Garwin said the standard airport runway for international flights is 12,000 feet long, about two and a half miles. A floating airport could have runways as long as anyone wanted them.

The first model anyone thought of was a huge aircraft carrier, but they are small in comparison to land-based airports, and no one is going to land a Dreamliner on one. Because of the size, the structural integrity would have to be much greater than that of a ship. It also would have to be more stable and keep its position.

Garwin's concept, based on a design originating at the Scripps Oceanography Institute, would be to build an island a mile wide and several miles long to accommodate the long runway. It would be floated out to sea in modules, 200 feet long, then flipped by flooding a section, something like how oil rigs are put in place now. The platform holding the runway and terminal would sit high above the water so that waves in storms would not wash over it.

Essentially, the deck structure would be supported on submerged floatation chambers anchored by taut cables to mass anchors 200–600 feet deep. If the buoyancy chamber is located deep enough to be below the base of waves it would be immune from heaving of seas. The trick, Garwin said, is the taut cables. An advantage of Garwin's design is that no power is needed to keep the deck steady. It would have large positive buoyancy but is kept restrained by the cables. The cables would also make sure that the platform doesn't move horizontally.

The study and the report that followed imagined such an airport in the New York Metropolitan area, where air traffic congestion is extremely high. They proposed an airport 12,000 by 5,600 feet with a surface area of 1,000 acres. The deck would have expansion joints made up of 200 by 200 feet modules. The passenger terminal and cargo area would be belowdecks.[11]

The only unsolved problem, Garwin reported, was getting people and cargo the twenty miles to and from land. He now thinks a tunnel would be the best way.

Storms, even hurricanes and forty-foot waves, would not impede operation. If even greater storms threatened, the deck could be raised and the chambers lowered. The structure could also resist tides and currents. Between the wire tension and five-foot thick cement on the runways there would be a stable platform.

"When you load a ship, the ship sinks deeper in the water," Garwin said. "But when you load my airport it doesn't sink deeper in the water because it's held at given height by the tension of the tables below. . . . No floating airport has been built on that design," he added.[12] Some tension leg platforms, however, are, and the system is a Garwin invention.

In the 1890s, a pair of French writers, Georges Le Faure and Henri de Graffigny, published a series of science fiction books with the strange name *The Extraordinary Adventures of a Russian Scientist* that featured men traveling around the inner planets of the solar system in a hollow sphere attached to a huge reflecting dish. The dish caught sunlight, and the sunlight propelled the spaceship. It may have sounded like an unlikely proposition at the time, but while the books came out, James Clerk Maxwell was developing equations showing that light, traveling photons, produced pressure, much as the wind does. In 1900, a Russian scientist, Peter Lebedev, actually measured such a phenomenon, and the notion was embedded in the Russian scientific psyche.

Enter Konstantin Tsiolkovsky, a weird, reclusive, self-taught scientist, living in an isolated village 200 kilometers from Moscow. He, of course, read a lot, including the fiction of Jules Verne, and may well have read Le Faure and de Graffigny. By 1920, he teamed up with a Latvian, Fredrik Arturovich Tsander, to propose a sun-driven spacecraft. Their idea was to deploy "tremendous mirrors of very thin sheets" to "use the pressure of sunlight to attain cosmic velocities."[13] Tsander eventually moved to Moscow, and he and Tsiolkovsky—who became the Father of Russian astronautics—were the leaders in the movement to develop space travel that enthralled Russian scientists and eventually led to Sputnik.

In 1951, Carl Wiley published "Clipper Ships of Space," in *Astounding Science Fiction*, which Garwin believes is the first description, at least outside of the Soviet Union. Tsander and Tsiolkovsky's work also triggered the imagination of Western science fiction writers. Their designs were the basis for the Sunjammer in Arthur C. Clarke's short story of the same name. Other writers, such as Jack Vance and Poul Anderson, also picked up on the idea.[14]

The idea lay dormant, particularly in the West, until Garwin, then at IBM's Watson Laboratory at Columbia, got involved. In 1956 he submitted a paper to the new IBM *Journal of Research and Development*, but it was rejected by the editorial board for not being serious enough. In 1958 he published a paper, "Solar Sailing—A Practical Method of Propulsion within the Solar System," which called for using off-the-shelf plastic film as "a solar radiation pressure sail." He was the first to do the math, the first technical analysis in the literature. A sail would, he wrote, be considerably cheaper than anything else proposed for space travel and more powerful. It was the first paper on solar sailing in a scientific publication.[15]

The key, as Garwin understood, was that a solar sail would require no fuel once it was launched into orbit. Fuel added weight to the spacecraft, greatly limiting its size, greatly increasing expense, and limiting the mission. The spaceship would be sent into orbit using conventional rockets, and "expelled" from its packing container, which he said would take eighty seconds, more than enough time to work with the ninety-minute period of orbit. The sail would be attached to the spacecraft by ribbons of the same material. It would gain altitude by furling the sail as the spaceship was approaching the sun in its orbit around the Earth, and unfurling it while receding.

He figured out how much energy would be required to manipulate the sail, the angle the sail should be set, and predicted such a vessel could go from the Earth to Venus in less than a year, much quicker than if it was assisted by a conventional rocket. Improvements in the sail might let the spaceship make the trip in less than a month. The paper was full of equations and calculations, putting actual science into the dream.

"It is obvious that there are considerable difficulties connected with space travel," Garwin wrote, "but those connected with the sail appear relatively small."[16]

The idea was revived in 1976, when Bruce Murray, head of the Jet Propulsion Laboratory, proposed building a solar sail vessel to rendezvous with Halley's Comet, which was to approach the Earth in 1986. Budget cuts and NASA inertia scuttled the mission. Private funding took over. In 1980, Murray, along with the astronomers Carl Sagan and Louis Friedman, formed the Planetary Society, the same year Kim Drexler of MIT patented what he called a Lightsail.

In 2009, the Planetary Society announced a private donor had contributed enough money to launch a Lightsail into orbit and eventually into deep space. Much of the funding efforts had been driven by Ann Druyan, Sagan's widow.[17]

The Planetary Society, using a heavy-lift rocket from the commercial SpaceX Corporation, launched a Lightsail in May 2014 for a test to make sure the sails would deploy properly, and was scheduled to conduct an actual Lightsail flight in 2017, all privately funded.[18]

Garwin thinks the solar sail would work for travel in the solar system, not beyond.[19] But others note that because the solar pressure was steady and stars emit light photons as well, it is possible to hit speeds of hundreds of thousands of miles an hour, and perhaps sail to nearby stars in 100 years—all without fuel and within the lifetime of two generations.

In April 2016 Breakthrough Initiatives announced a $100 million effort to use light-propelled craft to travel in a century from Earth to the star Alpha Centauri, a distance of four light years; the light pressure driving the craft would come from lasers or Earth or in orbit around the Earth rather than directly from the sun.

Here's a conundrum. Most of the world's electricity supply comes from coal. The places that need electricity the most are large cities and factories, almost all of them far from the coal mines. The coal must be either transported to the power stations by rail or barge, or, if the power station is not near the coal mine, there must be an efficient way of transmitting the power to where it is needed. Either way is costly.

This was particularly true in Appalachia in the 1960s. PSAC decided to see what technology might alleviate the problem. Garwin's solution was superconducting transmission lines—lines so effective it would make it economically feasible to transmit electricity over long distances using extremely low temperature to boost the efficiency of the lines.

"We wanted to see what could be done in the current state of the art," Garwin said.[20]

The result was a 1961 Garwin think piece for PSAC and a 1967 journal paper that is something of a classic in the field, "Superconducting Lines for the Transmission of Large Amounts of Electrical Power over Great Distances," written by Garwin and Juri Matisoo, an IBM engineer for the Proceedings of IEEE. It would permit construction of nuclear or solar power plants to be near oceans or deserts and still get massive amounts of electricity to where it is needed.

What they came up with was enormous, 100 gigawatts of direct current at a distance of 1,000 kilometers or 621 miles, about half the electric power then generated in the United States.[21] They chose direct current because alternating current would make the wires shake too much. The lines would be buried in the ground. The only loss in the transmission would be the power needed to maintain the cables at four degrees kelvin (−452°F). The wires would be made of a combination of niobium and tin or niobium and zirconium all surrounded by a liquid nitrogen shield. They reported that the line would likely pay for itself in three months and save $1 billion a year in transmission costs even with the cost of refrigeration, less than one percent. It would produce less pollution.

"This is not an engineering study," the men wrote, "it is the considered option of one person [Garwin], and the result of some approximately 100 hours of direct effort on the problem. The main question is whether one wants to make the risks of interruption inherent in having a large portion of the nation's power transported on a single line. . . . If two things are feasible then the comparison between them must be made on the basis of cost." The superconductive lines would be cheaper. Whether it was desirable or necessary was a different matter.[22]

Advances have been made in superconductivity, raising the temperature at which the phenomenon occurs, but the Garwin-Matisoo line has yet to be built.

During World War II, the military had a weird yet effective way of getting mail, light freight, small packages, and even an odd human to or from remote encampments where roads were nonexistent and planes couldn't land. A plane would go through a maneuver adapted from airplane races and use ropes. Garwin, one day in 1975, working on a JASON project, published a paper presenting a cheap and easy way to refuel aircraft in flight without huge investment in tanker aircraft, an adaption of the old technique.

The US Air Force, he said, is proud of its ability to refuel aircraft, greatly extending their range.[23] It has a whole fleet of tanker aircraft outfitted with in-flight refueling booms, and it trains pilots of combat planes to attach themselves to the tanker and transfer tens of thousands of pounds of fuel while they are flying at altitude hooked together. It is costly and, because it requires precise control, dangerous. It is made more complicated by the fact that since airplanes can't stay up forever, a large fleet of tankers is needed at great cost. But the fleet makes it possible for B-52 bombers from the Midwest to make bombing runs in Afghanistan and return to their bases without ever touching the ground.

It is just the sort of problem Garwin loved to find solutions for that were cheaper, easier, and ever so slightly off the wall. It's not clear he ever thought they'd be adopted. This scheme was called DSTAR, Direct Sea to Air Refueling.

DSTAR would base tanker ships at sea. They would be cheap and, Garwin said, didn't have to be very good tanker ships.[24] An airplane could come along to a low altitude, circle the ship, and have its drink there rather than at a higher altitude. It would be made possible by the planes performing a pylon turn, a steep, banking turn around a fixed point on the surface, just as airplane racers once did.

The plane would drop toward the tanker, snag a thin cable, and set up its pylon turn, along with spewing out a hose line. At first the line follows the airplane around, but after the plane uncoils enough, gravity would take over and the line would go straight down the center of the turn, where it would be captured by the tanker. It would depend on the correct speed, altitude, and the angle of the plane's bank as it flew its circle. The line would remain vertical and stationary.

During the war, someone would grab the line and attach a mailbag, and it would get pulled up to the plane. Mail also could go down the line during a pylon turn. It was called a long-line maneuver, and Garwin said even pilots in the 1920s and 1930s employed such a technique.

The clever addition Garwin added was that if the line was long enough, centripetal force on the hose line would force the fuel to go up the hose and out the open end in the plane's fuel tank. Pumping would be unnecessary as long as the end in the plane was open.[25]

"The standby costs of the DSTAR ship are very small," he wrote Josh Lederberg.[26]

Garwin's paper drew little attention. In a letter to John J. Martin, assistant secretary of the US Air Force, he said the reaction in the Pentagon was dismissive. "Air Force appears to believe that the tanker fleet (KC-135 and the new KC-10) is a feel good, while buying DSTAR ships would cost money. I do not propose that there be a single DSTAR ship, but that there be quite a few, positioned at many places in the ocean in order to allow for flexibility to respond to weather and tactical needs."[27]

He thought the amount of time any plane could drink from the tanker should be limited to five minutes so that an entire fleet of aircraft could refuel at the same time. The air force also had qualms about the weight of the fuel flowing to the "low wing" as the plane banks in its turn. Garwin said they still do not understand the physics. "An aircraft in a bank does not have a 'low wing' and one does not need 'sophisticated fuel systems' to compensate for the tendency of the fuel to run into the 'low wing.'" Autopilots could easily control the pylon turn, rejecting another air force objection.[28]

Two years later, after DSTAR seemed to be going nowhere, Garwin wrote to Harold Brown, then secretary of defense. Brown had raised the question of cost and implied the US could not afford to switch entirely to ship-based refueling. Garwin felt that if a war lasted long enough there would be money for both DSTAR and conventional refueling vessels.

"The question might be whether people are going to die (in relatively small numbers and with small probability) in refueling maneuvers, or for lack of support on the battlefield," he wrote Brown. He pointed out the advantages of DSTAR, its flexibility and safety. Then he suggested the Pentagon was not paying attention. It was an old story of a military service that didn't want to hear something that went against the status quo and its self-image.[29]

"I just want to point out to you that the amount of work which has been done in the analysis of this concept is totally inconsistent with the possible benefit. Certainly there is a possibility that detailed analysis and experiment will show that the system is not sufficiently safe, or that it is too costly. But the job of the Air Force is not to have fun or to buy airplanes—it is to support U.S. political and military goals. To do this, we must ask how we can get the greatest capability at the earliest time, within a given budget allocation," Garwin wrote.[30]

He tried again with Zbigniew Brzezinski ("Dear Zbig"), the national security advisor. "What is ironical to me in the face of this accepted dependence on imported oil is the nation's inability to have the air force depend even a little bit on the navy for operating the

DSTAR ships (or to make the organizational changes within the air force which would allow it to design, procure, and operate DSTAR stations)," he wrote.[31]

Nothing came of it.

But twenty years later, Lederberg met an admiral who admitted he had been briefed on the notion years before and thought it was a good idea. The admiral, James Watkins, chief of naval operations—the boss of the uniformed navy—said, however, that he thought DSTAR was technically feasible but would run into "socio-political obstacles."[32]

It was never heard from again.

CHAPTER EIGHTEEN
RUMPLED

As he pushes ninety, Garwin does not look his age. When he is doing official business, he will wear a tie, and he dresses up for JASON business. But normally he is in chinos. He is rumpled. His white hair looks like it has not been attended to for a while. If his pictures are any indication, it's been years. He has been mistaken for Bernie Sanders. He has the look of the absentminded, which is most surely not the case. He is serious most of the time but has a sense of humor hidden beneath the gruff manner and can, as many have found out, be devastating.

He and Lois live in an upscale apartment building in Scarsdale, New York, within commuting distance of his IBM office where he still spends time. He is fully electronic. He describes most of his furniture as "Danish modern," or "midcentury," except for a mirrored cabinet he made himself. The Danish modern actually did come from Copenhagen. He has no hobbies. He travels to Washington frequently, and still flies off to places like China on a regular basis to discuss disarmament. He has acquaintances and colleagues around the world in the hundreds. He has collected almost everything he has written or has been written about him, or to him.

After volunteering for decades at an Alzheimer's facility, Lois now has the disease herself. As of this writing, they are still traveling.

The Garwins are not very religious. The family was twice-a-year Jews, going to their synagogue during the High Holidays. On the other hand, they are founding members of a local Reform Jewish synagogue, the Westchester Reform Temple. When the synagogue was formed, like many others in America, it held its first meetings in a church. It now is a huge synagogue with 1,200 family units. The

Garwins celebrated Hanukkah at home, Jeffrey said.[1] Garwin can parse out the Hebrew alphabet but does not understand Hebrew.

By every conventional standard, his children turned out successfully. Two of the three—Laura and Jeffrey—have PhDs, Jeffrey from Yale with an undergraduate degree from Stanford. He also has an MD. Tom is ABD—everything for a PhD except a dissertation—from Harvard's Kennedy School of Government.

Laura's doctorate is in earth sciences from Cambridge University, with undergraduate degrees from Harvard and Oxford. She was a Rhodes Scholar. She worked at the science magazine *Nature* in its London office, first as a subeditor, then manuscript editor, in the physical sciences team, and finally in charge of the team as the physical sciences editor. She moved to *Nature*'s Washington office as North American editor in 1996, then returned to Harvard in 2001 as its executive director of the Bauer Center for Genomics Research. Her heart, however, was in music. She practiced the trumpet before she went to work and then after—first thing in the morning and again at night.

"It became clear to me, as it should have been clear all along, that life was finite. . . . If there was something I wanted to do with my life, I should probably get on and do it," she says.[2]

Her parents supported the move, although her father asked about her supporting herself. To her surprise her colleagues in science and at Harvard were a bit envious she would dare. She sold her car, and moved to London to study at the Royal College of Music and is now a professional trumpet player in London, freelancing, and is principal trumpet of the Orchestra of St. Paul's, a chamber orchestra based in Covent Garden.

Following his PhD in biochemistry and MD from Yale and a postdoctoral fellowship at Harvard Medical School, Jeffrey began his career in biomedical development as a senior scientist at Biogen. He subsequently served as a medical director for pharmaceutical companies (prescription and OTC) and as an executive and consultant for medical device companies. Over the course of his biomedical research and development career, he was an author of twelve patents and significant PCT applications and nineteen scientific papers. His most commercially successful patent covered the production of Eggland's Best eggs, where he served as the executive overseeing clinical development and product quality assurance. Eggland's Best was the first national brand of "designer" eggs. Another commercial success was the combination of loperamide and simethicone (Imodium Multi-Symptom). Jeffrey has spent much of his career in the overlap between biomedical development and regulatory affairs, and is currently a regulatory medical writer in North Carolina.[3]

Tom has worked in the Office of the Secretary of Defense, worked as a senior professional staff member of the House Armed Services Committee, and has done contract work for the Intelligence Community, Nuclear National Security Administration of the Department of Energy, as well as the Department of Defense. He has published articles on defense strategy, arms control verification, and counter-terrorism. After the Bill and Melinda Gates Foundation launched a national search to find someone to set up a foundation-wide office of Impact Planning and Improvement, it selected Tom as a director. He had earlier worked for the John D. and Catherine MacArthur Foundation, helping to set strategy for a new Program for Peace and International Security, and more recently has consulted for the Rockefeller Foundation.

"You know, it's not a lot of kissing and hugging," Lois said. "[Dick's] mother was a wonderful mother, but it wasn't a very demonstrative family. We've become more demonstrative as we've aged. I think we were the kind of parents who wanted to be sure that everything was done just right. Not so much that the children did everything just right, but that we did everything just right for the children."[4]

Garwin obviously could not discuss work with his children. Occasionally there would be an oblique reference. His children knew that Los Alamos did nuclear weapons, but Tom said it never really sank in what that meant.

"I think I was probably the youngest kid in the block to understand what classified work was," Jeffrey said. "Namely, I didn't know any-thing about [his work] and was never going to know anything about it. . . . I would know after the fact or after something was published, Dad would occasionally say 'oh yeah we've been working on this project for a while and we just, you know, released it' . . . all the antisubmarine warfare stuff which apparently he's been doing forever. I was aware that he worked on antisubmarine warfare issues, but I still don't know anything about what he did."[5]

When Garwin won the R. V. Jones Intelligence Award for his work on spy satellites, it was something of a revelation. Jeffrey said, "Before that, the only thing I remember is that he would occasionally come up with information about how spy cameras worked. And I would think that was pretty neat. How'd he know that? . . . But he's a physicist, so he knows everything. But he never let on as to what he was doing. He would sometimes mention that he was going to a meeting, and there was a committee, and he would mention some of the names of the people on the committee, but as far as I knew, they were just regular old physicists."

Jeffrey figured out some of what his father did with the hydrogen bomb when Garwin's battles with Teller became known. He has been under the impression since that Garwin just did the trigger for the bomb, but he did more than that. Laura said she knew he did some work on the bomb, but it wasn't until a *New York Times* article when she realized how critical it was.

Lois Garwin had a hint before then, however. At a science meeting at Erice, on the western tip of Sicily, an international conference that is a regular stop for the Garwins, a discussion broke out about disarmament. She does not remember the year, but Garwin was at the point when he wanted to either reduce or eliminate all nuclear weapons. Teller and his allies were there—not unusual. They were, Lois said, "gung-ho. . . ."[6]

"That was the meeting at which Edward said, 'Well, if there's a third world war, you'll be responsible,' to Dick. It was also at that meeting that he divulged that Dick had had a seminal . . . contribution to the development. . . . I was so thunderstruck by his saying that in public that I really don't remember why he said it. It may have been in the back of my mind I thought that maybe it was because he—these people were really ganging up on Dick, and he felt he needed to say something good about him. Or, whether he was trying to discredit Dick by saying, here's a guy who developed or made a very important contribution to the development of the hydrogen bomb, and now he wants to get rid of it?"

Teller felt that Garwin's opposition to the Star Wars antimissile defense system was anti-technology and countered by character-izing Garwin as wedded to offensive nuclear weapons rather than the appealing concept of a perfect defense.

When his role in history was later revealed, people who had known Garwin for a long time were stunned. They were amazed Garwin had never said anything.

William Perry, former secretary of defense, said that Garwin could be an "acquired taste. I liked him. I certainly respected him. He is very smart and sometimes a little impatient with anyone not quite as smart as he is. . . . I liked him a lot."[7]

His impatience could awe those around him. Physicist Jonathan Katz, a fellow JASON, remembered one incident. A group of scien-tists were stymied in an experiment. "And they said, I don't really know what happens here. Just we don't quite understand this. So Dick looked at it, and he said, 'I think your diffusion pumps stalled.' And these guys looked at each other and thought for a moment, or maybe more than a moment, and they turned to him and said, 'You know, you're prob-ably right. That's probably what happened.' And it was just spectac-ular. Here he had only the roughest inkling—the roughest description of their experiment—yet he was able to see, to understand what was going on and to diagnose a problem that these people had come across weeks or months—probably months earlier. Of course they were the people who built the experiment and sweated over it day after day, and they hadn't been able to figure out what was going on. And he, with his marvelous flash of insight, having heard these people talk for half an hour or an hour, probably having seen these particular results for thirty seconds or two minutes or something like that, immediately realized what the problem was. . . . Just truly spectacular."[8]

Murph Goldberger, whose friendship with Garwin went back to the school days, said that Garwin was not a great role model because no one can keep up with him. He probably "depresses the hell out of them," he said of young physicists. "They're young hotshots and he knows more than they do."[9]

"He can be impatient with people who don't pick things up quickly," Katz said. "That certainly has happened. He got into some big public controversies about ballistic missile defense a couple of decades ago. Some of it was a policy disagreement, and some of it was a certain degree of impatience with people—not scientists generally—but government bureaucrats who didn't quite appreciate some of the technical issues involved. When your mind goes faster than just about anyone else's, it's quite natural to get a little impatient occasionally."[10]

"I remember visiting somebody else's office with Henry Kendall, circa 1970, and the Xerox machine was off-kilter or something," said colleague John Cornwall, "and he said, 'Oh, let me take a look.' [A sign] says, 'Don't Open, Don't Touch.' He takes out his Swiss Army knife. And these people were horrified that someone would dare touch something as sacred as the innards of a Xerox machine. He said, 'Oh, it's this device. It's bent.' And now it's bent the right way."[11]

"Somebody held up a piece of machinery," said astrophysicist Mal Ruderman, "and Dick immediately said, 'Oh that's the so-and-so of a such-and-such of a Xerox machine.'"[12]

One time Garwin was in Geneva at a treaty negotiation, visiting Herb and Sybil York who were living in a beautiful mansion that the US government owned or leased. As York related to Dan Ford, it was "a beautiful, beautiful place but with a creaky old heating system." The place was cold. When they went down to the basement to check out the heating controls, they saw a sign warning against anyone touching anything. The furnace turned out to be an old one with pipes and valves going in and out. Garwin looked at it, turned a few valves, and there was heat.[13]

Sometimes his fiddling didn't work out quite as expected.

"I'll give you an example," Ed Frieman said. "I tease Dick all the time. He came to our house, when we had a big house up on the hill here. We had a swimming pool, and in the swimming pool is one of those crawlers that goes around and sweeps it all up. And it wasn't working. So Dick [yanks it]. I said, 'Dick, leave it alone.' No. He grabs it, yanks it out of the water. I mean the thing—he broke it. He can't resist it. He's just drawn to these things."[14]

"And he can be very intimidating to these people who come and brief us," said Cornwall. "In the days when people used viewgraphs, they don't use them anymore, Dick would walk in late, and he would take the viewgraphs that the guy had shown and the ones he was going to show, and he would flick through them all, and make an instantaneous decision as to whether he was going to stay or not, and then put them back. And often when he did stay, he would begin to point out many deficiencies and errors in the briefer's presentation. It could be very stressful for someone who is not really well-prepared."[15]

Garwin's people skills are somewhat legendary. According to Jack Ruina, at one PSAC panel, an admiral got up to talk, and four sentences into the admiral's presentation, Garwin informed him that his presence wasn't Garwin's idea. He wanted to hear someone who knew what he was talking about. Ruina said he didn't quite put it that way, but the message was clear. "You don't talk that way to anybody, no less an admiral in charge of anything," Ruina said.[16]

In fact, he is famous in Washington for tearing apart admirals and humiliating generals. In the JASON music skit at its anniversary part, one character sings to the tune of "Just you wait Henry Higgins: Oh, Richard Garwin/ Just you wait until your clearance is reviewed/ Oh Richard Garwin/ Do you think it is really going to be renewed?" It was.[17]

His memory even at his advanced years is prodigious. While it is not eidetic or photographic (he doesn't remember everything), names, dates, events, conversations, and facts flow instantly. He is almost never wrong.

"So what else can I say about Garwin? I could say that he wouldn't be a person I would use for political judgment," Ruina told Dan Ford. "He has a tendency—this is very clear, and I'm sure you've heard this from everybody—he has a tendency to have technical solutions to the world's problems. Technological or some kind of technical-type fix. If Garwin was here and we had a long conversation about my difficulty with my cousin, he would suggest solutions that would be technical solutions, like 'Why don't you write her every third week on Tuesday night.' Some kind of gadgetry, maybe it's social gadgetry, solution to the problem. I've heard him do that in social situations where the solution was—to exaggerate—if the government was doing a program which was absolutely wrong in every way, based on what the technology has to offer, and what the policy implications are, Dick would be very negative about it, because he had a better solution. Rather than saying the whole thing's crazy, [he'd say] 'We don't need that at all.'"[18]

"Technically, he is very good and we love him for that," said Freeman Dyson, also a JASON and a sometime critic.[19] His treatment of Joe Weber with his gravity waves, however, was a good example of the hard edge and Garwin's impatience with perceived imperfection. And perceived cruelty.

One of his closest allies was Wolfgang "Pief" Panofsky, a Stanford physicist. "We are good friends, we talk shop often on arms control issues. He tends to emphasize technological solutions to what others may consider to be fairly basic human problems. I mean, he is basically a real technocrat at heart," Panofsky said.[20]

"He was a member of the famous Rumsfeld Commission in the late 1990s, which was charged in 1998 with assessing the threat to the United States of ballistic missiles. But to most observers, [it] had a not so hidden agenda of justifying ballistic missile defense. Now Garwin basically did an honest job on that committee of basically assessing the North Korean threat and in turn, the fact that he who is usually more identified with the arms control side of things and who was usually very much identified with the loyal opposition to ballistic missile defense, nevertheless found that his concurrence with the findings of the Rumsfeld Commission and the subsequent use of that report to justify BMD."

Garwin was accused at the time of being naïve, of being oblivious to the political implications while still doing "an absolutely straightforward honest job on the technological parts of the thing," Panofsky said. "I mean, that's one example where he basically proceeded to analyze the explicit task given to that Commission but . . . since that Commission was not asked to recommend specific remedies to the ballistic missile threat, but only to analyze the ballistic missile threat, he kept to that."

Garwin typically analyzes the technical issues in isolation on their own merit, without looking at what other people will do with his findings. Again, the man at the guillotine. "Dick Garwin is a unique phenomenon and the fact that he separates those two compartments gives him a great deal of strength because he has got very sensible views on the insanity of the nuclear arms race or the escalatory nature of ballistic missile defense and all these things," Panofsky added. "But since he is willing to look at the technical merit of things on its own, people trust his work whether they are hawks, doves, or tame hawks . . . whether they are tame hawks, wild hawks, tame doves, or wild doves. I mean, taking those four categories of animals, they all trust him."

One JASON told Ann Finkbeiner he thought Garwin was "the most informed person in the United States on defense problems."[21]

"He's more than my ally," said Ted Postol, who has been a gadfly to the government for years. "He has been more of my protector and mentor than an ally. Because he is right."[22]

Some of his old friends, like Walter Munk, believe he has mellowed with age. Munk describes him as "caring."[23] Lois's Alzheimer's undoubtedly has helped bring that out. As for Garwin, he once said he did not agree with Plato, quoting Socrates, that an unexamined life is not worth living. "I like technical work," he told Ann Finkbeiner. "It really helps to do the calculations yourself, to be intimately familiar with the technology. So when you are the best person of the job, and it's a job that should be done, you do it. So that's what I do. I like to be helpful. I believe in progress. And I'm very good at taking the next step."[24]

"I don't like bureaucracy," Garwin said. "I am not particularly good at it. I worry more than people should who are managers and I have many other things to do. If I am not essential in such a role I try not to do it."[25]

Garwin's ally, friend, and sidekick for decades was Stanford physicist Sid Drell. They met when JASON was being formed in around 1960. Drell served on PSAC's Strategic Military Panel, which concentrated on America's deterrent force, and was chairman for two years. Drell specialized in the early years on designing satellites to spot Soviet missiles by tracing their infrared signatures as they rose over the horizon. When Corona satellites were launched, the first images were blurred by "beautiful but concealing patterns that looks like flames or brushes or flowers," Garwin said,[26] and Drell was called in to fix it. It turned out what was blocking Corona was a corona discharge that occurred when the film was being unwound from the spool by static charges. It was his initiation into the ultra-secret world of surveillance satellites.

When Henry Kissinger asked a group of scientists to look into new satellite systems for watching the Soviet Union, Garwin and Drell were the only members of the informal committee who had detailed knowledge of the options, from their work with the Land Panel on overhead reconnaissance. The two drafted a top-secret memorandum, hand delivered to Kissinger, recommending an electro-optical satellite. It was developed. The draft remains top secret.[27] Drell's work was especially important in finally establishing a Comprehensive Test Ban Treaty, Garwin said.

The two men, something of the last survivors of twentieth-century political science, are still in close contact, Garwin in Scarsdale, Drell at the Hoover Institution on the Stanford campus.

John Parmentola, retired head of nuclear power at General Atomics, first ran into Garwin as a graduate student at MIT working on a panel with Garwin on a study of particle beam weapons, weapons both men considered "ill-conceived."

"What I saw [at MIT] was an incredible display of skill and talent by the way he approached the whole thing," Parmentola said. "I grew up in the center for theoretical physics at MIT. Everywhere you go down the corridor everyone is a legend in physics. I was totally amazed by the knowledge he had and the skills he had and the ability to answer questions from the other participants with such great facility.[28]

"What it showed is how a scientist could use those skills to address major policy issues. I saw how Dick conducted himself. He became an example for the rest of us on how to conduct yourself as a scientist in the public policy arena. Dick always tried to adhere to the objective truth that science and technology provided, and to use that to infuse public policy. He was relentless in the way he did this. It showed that if you could use those skills properly and you did persist you could affect things and change the way policy is formulated and implemented. . . . He does require people [to] pay attention to facts and logic."

In 1977, Rae Goddell (now Rae Simpson), a doctoral journalism student at Stanford, wrote a dissertation (later a book) called "Visible Scientists," describing how some scientists get to be known in the public by popping up in the media regularly. The scientists she chose are largely out of the public eye now—most of them are dead, actually—but the way they became famous still holds today, even in the absence of social media that scientists in the 1970s could not have known about. It is hard to think of Linus Pauling, William Shockley, and Carl Sagan passing up Twitter.

To be visible in the media, a scientist needs several attributes, including an urge for self-promotion. Garwin appears to have none. The reason hardly anyone knew of his work with the hydrogen bomb until recently is that he never bothered to tell anyone, including Richard Rhodes who was writing the definitive history of the bomb. It took a Teller heart attack for the word to get out, and even then, few paid attention. You have read the first full public account in the sixty years since Mike outside of the scientific press.

A visible scientist must be accessible, able to condense what he or she has to say, and have some understanding of how reporters operate. Visible scientists should be tenured so their job isn't jeopardized by what they say, or at least have solid employment. They should be an expert on something the public wants to know about. Garwin is now retired from the right job and was always protected by his employer so he could speak fearlessly and did. Yet he is still largely unknown.

One of the most visible scientists of the end of the last century was the astronomer Carl Sagan. Sagan was not shy about self-promotion. He cultivated science writers, became friends with many, and was accessible to all even after his television program made him something of a celebrity. It helped that he was a nice man. He was not above leaking information he thought ought to get out, and once snuck the author of this book into a meeting at the Jet Propulsion Laboratory that was closed to the press so the author could have an exclusive story about a Mars landing for his newspaper. He did it for others as well, so complaints of favoritism were few. Many journalists had his telephone number (clearly before cell phones), and if there was a story to do and the writer needed something about space travel—piloted or otherwise—or about the possibility of alien life, and needed it on deadline, a conversation could begin this way: "Carl, I need two paragraphs on the virtues of a manned Mars program."

Sagan would give exactly two paragraphs. It would be accurate (if you understood his biases for space exploration) and contained exactly what the writer needed, no more, no less. It would be perfectly grammatical, always thoughtful, and occasionally lyrical. He was himself a great writer. If the reporter decided he needed more, Sagan, who seemed to understand the structure of news stories, would oblige sometimes in inverted pyramid style, the way most news stories are written. He got lots of calls. (The anthropologist Margaret Mead, also in Goddell's book, would write the story for you if you were young and inexperienced.)

Shockley, the coinventor of the transistor, known more publicly for his racist views,[29] could play the press like a pipe organ, guaranteeing news coverage by instigating protests and going over the head of reporters to their editors. Pauling, a two-time Nobelist, and his vitamin C research were irresistible. None of that applies to Garwin.

Richard Garwin would not have made Goddell's list despite his frequent presence in Washington and his influence, which is why he is the most influential scientist you never heard of. William Broad in the *New York Times*, probably the reporter Garwin knows best, called him "among the last living physicists who helped usher in the nuclear age."[30]

Garwin said that the Star Wars controversy required him to talk to the media. "Very often they are on deadline and they want a quote, so I learned early on, I asked for their questions in writing so I would have a record, and still do. I really needed to be able to show what I was asked and what I responded in order to avoid accusations of providing classified information," he explained.[31]

That won't work for most reporters on deadline.

Garwin also is a difficult interview. No simple question is likely to elicit a simple answer. He knows too much, and it must come out. One result is that there is no such thing as a quick interview with Garwin, a reason it is unlikely he would get many calls from reporters unless they had the time and were willing to—or need to—go into the weeds with him. Those who take the time would find him full of good stories enhanced with his superb memory. He can also be surprisingly frank, being beyond the point where he cares what others think of him—if he ever did. He is a polymath and knows things well beyond physics and loves talking about them.

His children understood this characteristic. He was always happy to help them with homework, although even they learned to put time limits on answers, as in "Dad, I only have fifteen minutes."

His daughter Laura recalled asking a question on a homework problem, and instead of just answering quickly, he gave two answers, one he described as "heuristic."[32] She was about twelve, and that was not a useful answer for her.

"It bothered my brother and sister," Tom said, "because they wanted help getting their homework done. He always had interesting things to say. Once he talked about differential equations when I had a question about algebra."[33]

Garwin also suffers fools badly, so the reporter better know what he or she is talking about. Another problem is that most of the things that interest him have to do with policy, and the American media doesn't do policy.

Many of his letters to the editors and the op-eds were never published but are in his online archives maintained by the Federation of American Scientists.[34] Sometimes he was simply pushing his opinion; sometimes he was protesting how he was quoted or edited. Since he was not shy of the media, he had ample opportunity.

In 1979, he was interviewed by Harry Reasoner for CBS's *60 Minutes* on his opposition to Star Wars, and particularly to charged particle beam weapons, which were to be used to down Soviet missiles. In a letter to producer Don Hewitt, he pointed out that Reasoner had interviewed him for eleven minutes, but the amount of the interview actually broadcast could be counted in seconds. He sent Hewitt the unedited transcript of the interview and what was broadcast, along with references to the articles he had written on the subject, pointing out that CBS had some of them before the interview. He assumed someone at CBS would have done extensive research. He invited Hewett to compare the two.[35]

In it he denied the weapons were controversial. Most scientists agreed they were not feasible. He rejected the idea that the Soviet Union had been spending the last twenty years trying to perfect such weapons, pointing out that if that was so, the Soviets were doing a poor job of it. He gave multiple reasons why the weapons would, even if they worked, not be the most effective method for downing the missiles.

The program contained two quotes from the eleven minutes and then cut to an interview with Teller. The letter to Hewitt was published in the *Bulletin of the Atomic Scientists.*[36]

Garwin can also organize scientists in politics for a cause and sometimes play the media. In the summer of 2015, when the Obama administration had worked out a deal with Iran to put some controls on its nuclear plans, twenty-nine scientists, including Nobel laureates, wrote a public letter to the president praising the plan despite considerable opposition. The ringleader of the letter? Richard Garwin, along with Rush D. Holt, a former congressman and head of the American Association for the Advancement of Science. Many of the signers of the letter had Q clearance, the highest security clearance possible in nuclear weapons. Many, like Frank von Hippel of Princeton, had been active through the years on disarmament matters as advisors to the White House and government agencies. Sid Drell, of course, signed. The thrust of the letter warned that Iran was "only a few weeks away" from having enough fuel for nuclear weapons, and it is likely the signers knew what they were talking about.

The letter began,

> As scientists and engineers with understanding of the physics and technology of nuclear power and of nuclear weapons, we congratulate you and your team on the successful completion of the negotiations in Vienna. We consider that the Joint Comprehensive Plan of Action (JCPOA) the United States and its partners negotiated with Iran will advance the cause of peace and security in the Middle East and can serve as a guidepost for future non-proliferation agreements.
>
> This is an innovative agreement, with much more stringent constraints than any previously negotiated non-proliferation framework.[37]

The only scientist invited to sign who refused was Steven Weinberg, a Nobel laureate at the University of Texas.

Weinberg wrote to Garwin that although he admired Garwin's judgment on military matters, arms control, and missile defense, "on this matter I can't go along with you." He said he did not share Garwin's "enthusiasm for the deal" as important respects this agreement falls short of goals previously announced by the Obama administration. Weinberg did not think the government of Iran could be trusted.[38]

"Your well-deserved prestige as a government adviser and defense intellectual will lead Congress and the public to give great weight to whatever you say on this issue," he wrote, and asked Garwin to reconsider.[39]

It made the *New York Times*, and the paper syndicated it around the world, and there were many stories on the Internet. For the first time, however, Garwin was widely acknowledged as a key designer of the hydrogen bomb, mostly because of Broad's story in the *New York Times*. He was identified as an expert on disarmament and defense.

CHAPTER NINETEEN

DECLINE OF INFLUENCE

The apotheosis of scientific influence in Washington was in the 1950s and 1960s, agreed William Perry, who was undersecretary of Defense for Research and Engineering and later secretary of defense under President Clinton.[1] Garwin was still a young man and had by then earned the reputation as the smartest man in the room—most any room. He was in great demand.

Scientists had come out of World War II as gods, Ted Postol said, and the Manhattan Project cloaked science in general, the physical sciences in particular, with an aura of glory.[2] They had produced a miracle, had ended the worst war in history, and showed what could be done if the government called on scientists and scientists—out of patriotism if nothing else—showed up. Have a problem? Call an expert.

The characters of President Eisenhower and President Kennedy played a role in the position of scientists. Eisenhower, a former general who served in World War II as the supreme commander in Europe, mistrusted the military. It was he, of all people, who warned Americans about the "military-industrial complex."[3] No Washington prophecy ever came truer. He apparently believed that science could solve many of the problems facing America and could make it immune from the pressures of politics and money. In that, he turned out to be wrong. Kennedy, with a degree from Harvard and the pedigree of new New England aristocracy, had the intellectual bias toward rationality, science, and fact. The two presidents did not hesitate to call on scientists when they thought science might be useful.

Eisenhower and Kennedy had two externalities to contend with. Perry named the Cold War and scientific and technological competition with the Soviet Union as reasons for the rise in science's prestige.

During the Cold War, he said, government leaders recognized they needed scientists and engineers, believing the survival of the United States was at risk. "It was life or death," Perry said. The threat was visible, missiles and nuclear weapons, and new weapons technology. The launch of Sputnik convinced Washington that the competition was real and potentially dire. Soviet technology now was a threat; only American technology could offer protection. There was nothing a foot soldier or a tank could do against an ICBM. When he was in the Pentagon, Perry said, he paid considerable attention to the Defense Science Board and went to its meetings as well as those of JASON.[4]

"When Dick was first on PSAC during the Eisenhower Administration, there actually was a crisis after Sputnik," said Frank von Hippel, a theoretical physicist at Princeton who has spent most of his career in policy issues. The administration was concentrating on nuclear weapons programs and defense.[5]

With the fall of the Soviet Union, America no longer sees itself in competition with anyone, Perry said.

Congress was generally out of the loop even when science was riding high. Garwin, in a talk at the Illinois Mathematics and Science Academy in 1993, said that congressional testimony on military procurement programs and weapons in general involved active-duty military officers and administration officials. When the Star Wars brouhaha broke out, it became clear that the scientists who built the nuclear weapons and radar during the war "could not be written off as ignorant about current topics and they were invited to testify." He called it a "sea change." Testifying to Congress became largely what Garwin did.[6]

And then things started to fall apart. Kennedy was assassinated, and Lyndon Johnson, decidedly not of Harvard, took over. He certainly did not dismiss science, but he just didn't seem to think it was as valuable as his predecessors. There was less emphasis on nuclear defense because he thought it a waste of money. The countermeasures were too easy to defeat, von Hippel said.[7]

The heroes of Los Alamos started to die off or retire. The first generation of science advisors had spent time in Los Alamos, and they were generally followed by their students. But the esteem wore off as generations replaced generations. The Vietnam War further greased the decline. Scientists had played an active role, from building McNamara's Wall to introducing Agent Orange as a herbicidal weapon on Vietnamese forests.

Two other factors played into the diminution of power: scientists became more specialized, and the government became larger and more decentralized. Where there once was Vannevar Bush who, in the 1940s, was the conduit for all of science in the government, now each agency has its own research and development department with scientists and public information officers. If you had any problem, there was a special person you could go see. The world got more divided.

Von Hippel said that in Washington, there is science for policy and policy for science. With policy for science, the government aims science in the direction it wants by providing support through the National Institutes of Health, NASA, the National Science Foundation, etc. In matters where science pushes policy, there is the Environmental Protection Agency, which issues regulations as science comes up with new findings. Each department or unit has its own science advisors and its own political agenda. There is no one place to apply pressure.[8]

Ted Postol added another factor, the influence of special interest groups, particularly with financial lobbyists who can insert their opinions that counter the advice of advisors and cripple the work of government scientists. The National Rifle Association can see to it that the Centers for Disease Control and Prevention is legally bound not to investigate the health impacts of gun violence. Congress can and did force the National Institutes of Health to fund research into alternative medicines and hinder science-based regulation such as those meant to limit climate change.

Some of it may be the fault of the scientific community.

"This is where Dick and I have disagreements, I think. I think the scientific community has been a major culprit for lots of reasons," Postol said. "The general public now sees experts in everything. Everybody declares themselves an expert. This is what I deal with all the time. They talk to some person who declares themselves an expert who really knows next to nothing because they don't know that they don't know."[9] The phenomena is well documented as the Dunning-Kruger effect or colloquially the "American Idol Effect"; people who have limited knowledge, or ability, of a topic can overestimate their prowess and consider themselves an expert.[10]

Cable news is one place where this happens frequently. The pseudo experts give out bad advice. And even real experts can be wrong. If the experts are wrong all the time, they should not be surprised to find that no one trusts them anymore, Postol said.[11]

Sometimes scientists really don't know, or they are genuinely split. During the crisis at Three Mile Island, when a nuclear reactor overheated, experts could not agree on what was happening and what, if anything, the public should do. Should people flee? Was it safe to remain? Reporters covering the incident had no idea what to tell the public or what to write. But the public insists on answers and met the indecision with anger.

There also is the influence of pressure groups, such as large corporations who "own fraudulent politicians" and pay off scientists who, Postol said, "prostitute themselves." A handful of scientists, funded by the fossil fuel industry, can give a patina of doubt about climate change using the same techniques tobacco companies used to deter regulation of cigarettes for generations even though the science is clear.[12]

The biggest change may be that the smartest people—people who would make great scientists—are going into the financial services business, especially people with degrees in physics and mathematics. Instead of producing wealth, they make their money manipulating other people's wealth and make great amounts of it. "It's an awful influence," Garwin agreed. "There's so much money to be made there."[13]

The primary example of higher level mathematics, partial differential equations and the like, to model economic markets was the Black-Scholes option pricing model. The equation was published in 1973, and Robert Merton used it to develop the model. It is used by hedge funds, and Black-Scholes is now ubiquitous in the industry.

Merton and Myron Scholes won a Nobel Prize in economics for their roles in creating the model. (Fisher Black, a mathematician, died before the award was given, and the prizes are not awarded posthumously). The positive industry response opened the door for others to use mathematics and statistics to get wealthy. The financial industry is capable of paying well for useful tools.

Garwin likes to say his competence as a government consultant is as an expert in technology. There were plenty of people around who could advise on legal and policy matters. "There were relatively few who had the experience and capability that I could bring to the table for the analysis of important technical issues," he said. He admits he frequently failed to move the bureaucracies as is true of most consultants who go against an agreed program. "There are many forces, both repulsive and attractive, that argue for various options. So it is difficult to get consensus on any one of them. Admiral Zumwalt himself felt that the massive aircraft carrier was obsolete in view of the potential of cruise missiles, but his alternative lost out because he did not stop the routine scheduled building of aircraft carriers in order to clear the way for alternatives."[14]

Garwin was initially bothered when he went into a government office and saw people sitting around reading newspapers. If they can't be working on their projects, why don't they multitask and do something else? It is of course possible that keeping themselves informed is part of their job.[15] He now says he understands most people don't have the freedom or the scope to be involved in many tasks at the same time.

In 2007, Garwin told a conference at Cornell of his frustrations, including his battle to get GPS funded. The report he made from the Air Traffic Control Panel for a satellite-based navigation system from PSAC was actually suppressed by the Federal Aviation Administration. Because it was never actually classified, the executive secretary of the panel could sneak a copy of the report to the national Technical Information Service to get the contents out. The role of science advisor in the government, Garwin said, was very weak at that time, and they got no help from the White House.[16]

Part of the problem, perhaps, is that Garwin does not have a close personal relationship with any of the people he advised. He did have dinner once with Ted Kennedy, but only because he met Stephen Breyer by accident on a long-delayed air shuttle and was invited to accompany him to the Kennedy home. Another disadvantage was that he was not at a university because he didn't want the endless stream of graduate students, and he got involved in so many different groups and projects.[17]

"Because my headquarters has been at the IBM laboratory, where I was isolated in this kind of activity, so one did not have political or executive branch personnel coming to talk to classes or to the policy-oriented groups that one finds at think tanks or universities. Sid Drell at Stanford and later the Hoover Institution at Stanford had an endless stream of people in his offices and a staff. He also hosts conferences, another way to plug into the establishment," Garwin said, perhaps with a bit of envy.[18]

He often has been a lone wolf.

AFTERWORD

The scientist who designed the most terrible of weapons does not really avoid thinking of the ethics and morality behind his act. It was an interesting scientific problem he solved, the man at the guillotine. But Garwin's life afterward has been devoted to seeing that his work on weapons does no harm to the innocent. He quotes the Golden Rule but interestingly, the Jewish version, paraphrasing Rabbi Hillel in the first century BCE, which is stated in the negative: "What is hateful to you, do not do to your fellow." Buddhists have a similar phrase.

"As for the ethics of scientists creating weapons about which irresponsible judgments may be made by the broader society," Garwin said. "I have taken the position that it is the responsibility of this broader society to decide, and mine to create such weapons but also explicitly to point out their problems, hazards, and the questions to be resolved."[1]

So far, humanity has decided and survived. For now.

In mid-November 2016, at the age of eighty-eight, just after returning from a consulting trip to China, Garwin was informed he had won the Presidential Medal of Freedom, the highest civilian award in the US. "The Presidential Medal of Freedom is . . . presented to individuals who have made especially meritorious contributions to the security or national interests of the United States, to world peace, or to cultural or other significant public or private endeavors," the White House announcement said. "From scientists, philanthropists, and public servants to activists, athletes, and artists, these . . . individuals have helped push America forward, inspiring millions of people around the world along the way."[2]

NOTES

CHAPTER ONE: THE TINKERERS

1. Richard L. Garwin, "The Secret Hans: For the Celebration of the 100th Birthday of Hans Bethe," *Aspen Center for Physics*, Aspen, CO, August 4, 2006, http://fas.org/rlg/rlg092006.html (accessed July 22, 2016).

2. Richard L. Garwin, in an interview with the author, May 20, 2014.

3. Richard L. Garwin, interview by Finn Aaserud, *Niels Bohr Library & Archives*, AIP, October 23, 1986, https://www.aip.org/history-programs/niels-bohr-library/oral-histories/4622-1 (accessed July 22, 2016).

4. Richard L. Garwin, in an autobiographical letter to Georges Charpak, November 7, 1995.

5. Garwin, interview with the author.

6. Ibid.

7. Marguerite Goldfarb, in an interview with Daniel Ford, October 2004.

8. Garwin, interview with the author.

9. Goldfarb, interview with Daniel Ford.

10. Garwin, interview with the author.

11. Ibid.

12. Garwin, autobiographical letter to Georges Charpak.

13. Ibid.

14. Garwin, interview with the author.

15. Ibid.

16. Richard L. Garwin, in an interview with Patrick McCray, June 7, 2001.

17. Ibid.

18. Lois Garwin, in an interview with Daniel Ford, July 1, 2004.

19. Ibid.

20. Garwin, autobiographical letter to Georges Charpak.

21. Lois Garwin, interview with Daniel Ford.

22. Ibid.

23. Garwin, interview with the author.

24. Ibid.

25. Garwin, autobiographical letter to Georges Charpak.

26. Garwin, interview with the author.

CHAPTER TWO: FERMI

1. Harris L. Mayer, *An Inconclusive Meeting of the Theoretical Megaton Group* (Los Alamos, NM: Los Alamos Historical Society, 2009).

2. Richard L. Garwin, in an interview with the author, March 28, 2014.

3. Ibid.

4. Richard L. Garwin, in an autobiographical letter to Georges Charpak, November 7, 1995.

5. Ibid.

6. Ibid.

7. Lois Garwin, in an interview with Daniel Ford, July 1, 2004.

8. Richard L. Garwin, in an interview with the author, May 20, 2014.

9. Richard L. Garwin, unpublished manuscript.

10. Lois Garwin, interview with Daniel Ford.

11. Jeffrey Garwin, in an interview with Daniel Ford, January 16, 2011.

12. Ibid.

13. Ibid.

14. Garwin, autobiographical letter to Georges Charpak.

15. Lois Garwin, interview with Daniel Ford.

16. Harry S. Truman, "Statement on the Hydrogen Bomb," January 31, 1950, transcript, *NuclearFiles.org*, http://www.nuclearfiles.org/menu/library/correspondence/truman-harry/corr_truman_1950-01-31.htm (accessed November 3, 2016).

CHAPTER THREE: THE SUPER

1. Kenneth W. Ford, *Building the H Bomb: A Personal History* (Singapore: World Scientific, 2015), p. 198.

2. Ann Finkbeiner, *The JASONS: The Secret History of Science's Postwar Elite* (New York: Penguin, 2006).

3. Richard Rhodes, *Making of the Atomic Bomb: 25th Anniversary Edition* (New York: Simon & Shuster, 2012), pp. 307–308.

4. Greg Canavan, in an interview with the author, May 25, 2016.

5. Rhodes, *Making of the Atomic Bomb*.

6. John Barbour, Associated Press, "Enrico Fermi's Daughter Has Clear Memory of Atomic Age's Dawning," *Los Angeles Times,* January 8, 1995, http://articles.latimes.com/1995-01-08/news/mn-17488_1_nella-fermi.

7. Jennet Conant, *109 East Palace* (New York: Simon & Schuster, 2005), p. 59.

8. Ibid., p. 55.

9. Finkbeiner, *JASONS*, p. 3.

10. Richard P. Feynman, *Surely You're Joking, Mr. Feynman!* (New York: W. W. Norton, 1985).

11. Finkbeiner, *JASONS*, p. 5.

12. Feynman, *Surely You're Joking*.

13. Kai Bird and Martin J. Sherwin, *American Prometheus: The Triumph and Tragedy of J. Robert Oppenheimer* (New York: Knopf Doubleday Publishing Group, 2007), p. 313.

14. Finkbeiner, *JASONS*.

15. Edward Teller and Judith Shoolery, *Memoirs: A Twentieth-Century Journey in Science and Politics* (Cambridge, MA: Perseus Publishing, 2001).

16. Conant, *109 East Palace*, p. 347.

17. Finkbeiner, *JASONS*, p. 7.

18. Richard L. Garwin and Sidney Drell, interview by Francis Slakey and Jennifer Ouellette, *Niels Bohr Library & Archives*, AIP, May 10, 2006, www.aip.org/history-programs/niels-bohr-library/oral-histories/30602 (accessed July 21, 2016).

19. Teller and Shoolery, *Memoirs*.

20. "Edward Teller (1908–2003)," *American Experience*, http://www.pbs.org/wgbh/amex/bomb/peopleevents/pandeAMEX73.html (accessed November 4, 2016).

21. Richard L. Garwin, "Working with Fermi at Chicago and Postwar Los Alamos," (lecture, Chicago, IL, 2001).

22. Zuoyue Wang, *In Sputnik's Shadow: The President's Science Advisory*

Committee and Cold War America (New Brunswick, NJ: Rutgers University Press, 2008).

23. Richard Rhodes, *Dark Sun: The Making of the Hydrogen Bomb* (New York: Simon & Shuster, 1995).

24. Ibid.

25. Ibid.

26. Ford, *Building the H Bomb*, p. 198.

27. Marvin "Murph" Goldberger, in an interview with Daniel Ford, June 2004.

28. Rhodes, *Dark Sun.*

29. Richard L. Garwin, in an interview with the author, February 12, 2016.

30. Richard L. Garwin, in an interview with Daniel Ford, June 27, 2004.

31. Richard L. Garwin, in an interview with the author, November 17, 2015.

32. Ford, *Building the H Bomb*, p. 198.

33. Robert Park, interview by W. Patrick McCray, *Niels Bohr Library & Archives*, AIP, June 7, 2001, https://www.aip.org/history-programs/niels -bohr-library/oral-histories/24351-2 (accessed July 22, 2016).

34. Ford, *Building the H Bomb.*

35. Rhodes, *Dark Sun*, p. 588.

36. Richard L. Garwin, in an interview with the author, May 20, 2014.

37. Ibid.

38. Richard L. Garwin, interview by Kenneth Ford, *Niels Bohr Library & Archives*, December 20, 2012, https://www.aip.org/history-programs/ niels-bohr-library/oral-histories/35680 (accessed July 22, 2016).

39. Ford, *Building the H Bomb.*

40. Rhodes, *Dark Sun.*

41. Garwin, interview with the author, May 20, 2014.

42. Rhodes, *Dark Sun.*

43. Edward Teller and Stanislav Ulam, *On Heterocatalytic Detonations I. Hydrodynamic Lenses and Radiation Mirrors* (Los Alamos National Laboratory, NM: March 9, 1951).

CHAPTER FOUR: GARWIN'S DESIGN

1. Greg Canavan, in an interview with the author, May 25, 2016.

2. Harris L. Mayer, *An Inconclusive Meeting of the Theoretical Megaton Group* (Los Alamos, NM: Los Alamos Historical Society, 2009).

3. Richard L. Garwin, "A Cruise Missile System for Europe," *Sack Memorial Lecture* (lecture, Cornell University, Ithaca, NY, September 26, 1978).

4. Michael Bernardin, in an interview with the author, May 25, 2016.

5. Ibid.

6. Richard L. Garwin, in an interview with the author, May 20, 2014.

7. Richard L. Garwin, interview by Kenneth Ford, *Niels Bohr Library & Archives*, December 20, 2012, https://www.aip.org/history-programs/niels-bohr-library/oral-histories/35680 (accessed July 22, 2016).

8. Richard L. Garwin, in an interview with the author, March 28, 2014.

9. Ibid.

10. Ibid.

11. Robert Park, interview by W. Patrick McCray, *Niels Bohr Library & Archives*, AIP, June 7, 2001, https://www.aip.org/history-programs/niels-bohr-library/oral-histories/24351-2 (accessed July 22, 2016).

12. Bernardin, interview with the author.

13. Garwin, interview with the author, May 20, 2014.

14. Ibid.

15. Mayer, *Inconclusive Meeting*.

16. Ibid.

17. Ibid.

18. Kenneth W. Ford, *Building the H Bomb, A Personal History* (Singapore: World Scientific, 2015), p. 198.

19. Garwin, interview by Kenneth Ford.

20. Canavan, interview with the author.

21. Garwin, interview with the author, May 20, 2014.

22. Ibid.

23. Richard Rhodes, *Dark Sun: The Making of the Hydrogen Bomb* (New York: Simon & Shuster, 1995), p. 588.

24. Garwin, interview by Kenneth Ford.

25. Rhodes, *Dark Sun*.

26. Ibid.

27. Ibid.

28. Garwin, interview by Kenneth Ford.

29. Michael Bernardin, in a discussion with Richard L. Garwin, November 19, 2016.

30. Richard L. Garwin, "Scientist, Citizen, and Government: Ethics in Action or Ethics Inaction" (lecture, Illinois Mathematics and Science Academy, Aurora, IL, May 4, 1993), http://fas.org/rlg/930504-imsa.htm (accessed October 19, 2016).

31. Bernardin, interview with the author.

32. Ibid.

33. Ford, *Building the H Bomb*, p. 155.

34. Garwin, interview by Kenneth Ford.

35. David Kestenbaum, "An Evening with Richard Garwin," *National Public Radio* (January 10, 2016).

36. Ibid.

37. Ibid.

CHAPTER FIVE: GARWIN, LEDERMAN, AND THE MARX BROTHERS

1. Lois Garwin, in an interview with Daniel Ford, July 1, 2004.

2. Murray Gell-Mann, in an interview with Daniel Ford, January 15, 2007.

3. Lois Garwin, interview with Daniel Ford.

4. Richard L. Garwin, in an interview with the author, May 20, 2014.

5. Richard L. Garwin, interview by Finn Aaserud, *Niels Bohr Library & Archives*, AIP, October 23, 1986, https://www.aip.org/history-programs/niels-bohr-library/oral-histories/4622-1 (accessed July 22, 2016).

6. Leon Lederman, interview by Roger Bingham, "Education, Politics, Einstein and Charm," transcript, *Science Studio* (March 4, 2008), http://the sciencenetwork.org/media/videos/2/Transcript.pdf (accessed July 22, 2016).

7. Harold Schmeck, "Basic Concept in Physics is Reported Upset in Tests," *New York Times*, January 16, 1957.

8. Richard L. Garwin, in an interview with the author, September 9, 2015.

9. Leon Lederman, interview by Roger Bingham, "Education, Politics,

Einstein and Charm," transcript, *Science Studio*, March 4, 2008, http://thesciencenetwork.org/media/videos/2/Transcript.pdf (accessed July 22, 2016).

10. Richard L. Garwin, "Fun with Muons, GPS, Radar, etc.," *Lee Historical Lecture* (lecture, Harvard University, Cambridge, MA, March 18, 2003), http://fas.org/rlg/FunWithMuons051607b.pdf (accessed July 22, 2016).

11. Richard P. Feynman, *Surely You're Joking, Mr. Feynman!* (New York: W. W. Norton, 1985).

12. Chen Ning Yang, "The Law of Parity Conservation and Other Symmetry Laws of Physics," (lecture, Stockholm, Sweden, December 11, 1957), https://www.nobelprize.org/nobel_prizes/physics/laureates/1957/yang-lecture.pdf (accessed November 6, 2016).

13. David Lindley, "Focus: Landmarks—Breaking the Mirror," *Physical Review*, 22 (December 2, 2008), p. 19.

14. Richard Hammond, *Chien-shiung Wu* (New York: Chelsea House, 2010).

15. Lindley, "Focus: Landmarks," p. 19.

16. Garwin, interview with the author, September 9, 2015.

17. Ibid.

18. Garwin, "Fun with Muons."

19. Ibid.

20. Garwin, interview with the author, September 9, 2015.

21. Ibid.

22. Richard L. Garwin, in an interview with Daniel Ford, June 26, 2004.

23. Ibid.

24. Lederman, interview by Roger Bingham.

25. Garwin, interview with Daniel Ford, June 26, 2004.

26. Ibid.

27. Lederman, interview by Roger Bingham.

28. Ibid.

29. Garwin, interview with Daniel Ford, June 26, 2004.

30. Ibid.

31. Lindley, "Focus: Landmarks," p. 19.

32. Gell-Mann, interview with Daniel Ford, January 15, 2007.

33. Ibid.

34. Garwin, interview with Daniel Ford, June 26, 2004.

35. Walter Munk and Ed Frieman, in an interview with Daniel Ford, July 2, 2004.

36. Norman F. Ramsey, in an interview with Daniel Ford, December 2004.

37. Garwin, "Fun with Muons."

CHAPTER SIX: IBM AND LAMP LIGHT

1. Richard L. Garwin, in an interview with the author, May 20, 2015.

2. Ibid.

3. Ibid.

4. Richard L. Garwin, in an interview with Daniel Ford, July 3, 2004.

5. Richard L. Garwin, in correspondence with the author, January 7, 2016.

6. Ibid.

7. Richard L. Garwin, interview by Finn Aaserud, *Niels Bohr Library & Archives*, AIP, October 23, 1986, https://www.aip.org/history-programs/niels-bohr-library/oral-histories/4622-1 (accessed July 22, 2016).

8. Ibid.; Garwin, interview with the author, May 20, 2015.

9. Garwin, interview with the author, May 20, 2015.

10. Garwin, interview by Aaserud; Garwin, interview with the author, May 20, 2015.

11. David Kestenbaum, "An Evening with Richard Garwin," *National Public Radio* (January 10, 2016).

12. Richard L. Garwin, "Scientist, Citizen, and Government: Ethics in Action or Ethics Inaction" (lecture, Illinois Mathematics and Science Academy, Aurora, IL, May 4, 1993), http://fas.org/rlg/930504-imsa.htm (accessed October 19, 2016).

13. Richard L. Garwin, in an interview with the author, April 28, 2016.

14. Richard L. Garwin, "Physics in the Interest of Society," *MIT Lecture Series* (lecture, MIT, Cambridge, MA, November 3, 2011), http://fas.org/rlg/110311%20PISp.pdf (accessed July 22, 2016).

15. Daniel Boslaugh, *When Computers Went to Sea: The Digitization of the United States Navy* (Los Alamitos, CA: Computer Society of America, 1999), pp. 117–25.

16. Jeremi Suri, "America's Search for a Technological Solution to the Arms Race: The Surprise Attack Conference of 1958 and a Challenge for 'Eisenhower Revisionists,'" *Diplomatic History*, 21, no. 3 (1997): 417–50.

17. Richard L. Garwin, in an interview with the author, March 29, 2016.

18. Office of the Historian, US Department of State, "*Report by the Technological Capabilities Panel of the Science Advisory Committee*," Washington, DC, February 14, 1955, https://history.state.gov/historicaldocuments/frus 1955-57v19/d9 (accessed July 21, 2016).

19. National Reconnaissance Office, "National Imperative (1945–1955)," n.d., http://www.nrojr.gov/teamrecon/res_his-NationalImp.html.

20. Ibid.

21. Boslaugh, *When Computers Went to Sea*.

22. National Reconnaissance Office, "National Imperative (1945–1955)."

23. Garwin, interview with the author, March 29, 2016.

24. Zuoyue Wang, *In Sputnik's Shadow: The President's Science Advisory Committee and Cold War America* (New Brunswick, NJ: Rutgers University Press, 2008).

25. Garwin, interview with the author, April 28, 2016.

26. National Reconnaissance Office, "National Imperative (1945–1955)"; Michel Fortmann and Albert Legault, *A Diplomacy of Hope: Canada and Disarmament, 1945–1988* (Montreal: McGill-Queen's University Press, 1992).

27. Garwin, interview with the author, April 28, 2016.

28. Ibid.

29. Ibid.

30. UN General Assembly, Thirteenth Session, Official Document, *Question of Measures to Prevent Surprise Attack*, statement from Soviet government provided in letter from Valerian Zorin (December 8, 1958).

31. Ibid.

32. Garwin, interview with the author, April 28, 2016.

33. Ibid.

34. Ibid.

35. Ibid.

36. Ibid.

37. Permanent Representative of the Union of Soviet Socialist Republics, "Letter to the Secretary General of the United Nations," January 16, 1959.

38. Garwin, interview with the author, April 28, 2016.

39. Garwin, interview by Aaserud.

40. Richard L. Garwin, in an interview with the author, May 6, 2016.

41. Garwin, interview with the author, April 28, 2016.

42. Ibid.

43. Wolfgang K. H. "Pief" Panofsky, in an interview with Daniel Ford, December 7, 2004.

44. Kestenbaum, "Evening with Richard Garwin."

45. Carl Kayser, in an interview with Daniel Ford, December 2004.

46. Jack Ruina, in an interview with Daniel Ford, December 2004.

47. Richard L. Garwin, in an interview with Daniel Ford, September 11, 2004.

48. Richard L. Garwin, interview by Kenneth Ford, *Niels Bohr Library & Archives*, December 20, 2012, https://www.aip.org/history-programs/niels-bohr-library/oral-histories/35680 (accessed July 22, 2016).

49. Ruina, interview with Daniel Ford.

50. Ibid.

51. Ibid.

52. Garwin, interview with Daniel Ford.

53. Ibid.

54. Richard L. Garwin, in an interview with the author, May 6, 2016.

55. Ibid.

56. Paul M. Doty, in an interview with Daniel Ford, December, 2004.

57. Ibid.

58. Garwin, interview by Finn Aaserud.

59. Garwin, interview with the author, May 6, 2016.

60. Ibid.

61. Garwin, interview by Finn Aaserud.

62. Garwin, interview with the author, May 6, 2016.

63. Ibid.

64. Garwin, interview by Finn Aaserud; Richard L. Garwin, "Fun with Muons, GPS, Radar, etc.," *Lee Historical Lecture* (lecture, Harvard University, Cambridge, MA, March 18, 2003), http://fas.org/rlg/FunWithMuons051607b.pdf (accessed July 2016).

65. Garwin, interview with the author, May 6, 2016.

66. Garwin, interview by Kenneth Ford.

67. Richard L. Garwin, in an interview with the author, June 24, 2016.

68. Ibid.

69. Ibid.

70. Ibid.

CHAPTER SEVEN: ADVISING PRESIDENTS—OR NOT

1. Richard L. Garwin, in an interview with the author, March 24, 2016.

2. Ibid.

3. Zuoyue Wang, *In Sputnik's Shadow: The President's Science Advisory Committee and Cold War America* (New Brunswick, NJ: Rutgers University Press, 2008).

4. Richard L. Garwin, "Physics in the Interest of Society," *MIT Lecture Series* (lecture, MIT, Cambridge, MA, November 3, 2011), http://fas.org/rlg/110311%20PISp.pdf (accessed July 22, 2016).

5. Wang, *Sputnik's Shadow.*

6. Ibid.

7. Ibid.

8. Ann Finkbeiner, *The JASONS: The Secret History of Science's Postwar Elite* (New York: Penguin, 2006).

9. Murray Gell-Mann, in an interview with Daniel Ford, January 15, 2007.

10. Richard L. Garwin, in personal document, "Sid Drell and National Security," undated.

11. Richard L. Garwin, "Physicists and Effective Public Policy: Science Matters," *New York Hall of Science*, New York, NY (June 14, 2012), https://fas.org/rlg/Physicists%20and%20Effective%20Public%20Policy.pdf (accessed July 21, 2016).

12. Garwin, interview with the author, March 24, 2016.

13. Ibid.

14. Ibid.

15. Ibid.

16. Ibid.

17. Richard L. Garwin, "Harold M. Agnew 1921–2013: A Biographical Memoir," National Academy of Sciences, 2015, p. 20, http://www.nasonline

.org/publications/biographical-memoirs/memoir-pdfs/agnew-harold.pdf (accessed November 7, 2016).

18. Garwin, interview with the author, March 24, 2016.

19. Ibid.

20. Zuoyue Wang, *Sputnik's Shadow*.

21. Richard Garwin, "How the Mighty Have Fallen," *Nature* 449 (October 4, 2007): 543.

22. Ibid.

23. Ibid.

CHAPTER EIGHT: JASONS

1. Ann Finkbeiner, *The JASONS: The Secret History of Science's Postwar Elite* (New York, NY: Penguin, 2006).

2. Ibid.

3. Murray Gell-Mann, in an interview with Daniel Ford, January 15, 2007.

4. Marvin "Murph" Goldberger, in an interview with Daniel Ford, December 20, 2012.

5. Finkbeiner, *JASONS*.

6. Gell-Mann, interview with Daniel Ford.

7. Richard L. Garwin, interview by Finn Aaserud, *Niels Bohr Library & Archives*, AIP, October 23, 1986, https://www.aip.org/history-programs/niels-bohr-library/oral-histories/4622-1 (accessed July 21, 2016).

8. Greg Canavan, in an interview with the author, May 25, 2016.

9. Richard L. Garwin, in an interview with the author, March 8, 2016.

10. Joel N. Shurkin, "The Secret War Over Bombing," *Philadelphia Inquirer*, February 4, 1973.

11. Canavan, interview with the author.

12. Walter Munk and Ed Frieman, in an interview with Daniel Ford, July 2, 2004.

13. Goldberger, interview with Daniel Ford.

14. Garwin, interview by Finn Aaserud.

15. Goldberger, interview with Daniel Ford.

16. Garwin, interview with the author, March 8, 2016.

17. Richard L. Garwin, "Fish Ragu (Fish, Radio-Receiving and Generally Useful)," *JASON Technical Note*, August 1981, http://fas.org/rlg/080081FISH%20Fish%20Ragu.pdf (accessed July 21, 2016).

18. Garwin, interview by Finn Aaserud.

19. Garwin, interview with the author, March 8, 2016.

20. Garwin, interview by Finn Aaserud.

21. Shurkin, "Secret War Over Bombing."

22. Walter Munk, interview by Finn Aaserud, *Niels Bohr Library & Archives*, AIP, June 30, 1986, https://www.aip.org/history-programs/niels -bohr-library/oral-histories/4790 (accessed July 21, 2016).

CHAPTER NINE: VIETNAM AND McNAMARA'S WALL

1. Richard L. Garwin, in an interview with the author, March 8, 2016.

2. Richard L. Garwin, interview by Finn Aaserud, *Niels Bohr Library & Archives*, AIP, June 24, 1991, https://www.aip.org/history-programs/ niels-bohr-library/oral-histories/5075 (accessed December 5, 2016).

3. Ibid.

4. Freeman Dyson, in an interview with the author, February 20, 2014.

5. Ann Finkbeiner, *The JASONS: The Secret History of Science's Postwar Elite* (New York, NY: Penguin, 2006), p. 90.

6. Office Secretary of Defense, Leslie Gelb, "Evolution of the War," *Pentagon Papers*, January 15, 1969, https://nara-media-001.s3.amazonaws .com/arcmedia/research/pentagon-papers/Pentagon-Papers-Part-IV-C-3 .pdf.

7. Finkbeiner, *JASONS*, p. 68.

8. Garwin, interview with the author.

9. David Kestenbaum, "An Evening with Richard Garwin," *National Public Radio* (January 10, 2016).

10. Murray Gell-Mann, in an interview with Daniel Ford, January 15, 2007.

11. Garwin, interview by Finn Aaserud.

12. Gell-Mann, interview with Daniel Ford.

13. Gell-Mann, interview with Daniel Ford.

14. Sarah Bridger, "The Best Intentions," *Slate*, September 4, 2015,

http://www.slate.com/articles/news_and_politics/history/2015/09/
hiroshima_nagasaki_hanoi_how_the_manhattan_project_generation_of
_scientists.html.

15. Garwin, interview with the author.

16. Finkbeiner, *JASONS*, pp. 67–68.

17. Garwin, interview with the author.

18. Finkbeiner, *JASONS*, p. 72.

19. Ibid., p. 79.

20. Garwin, interview with the author.

21. Ibid.

22. Garwin, interview by Finn Aaserud.

23. Garwin, interview with the author.

24. Finkbeiner, *JASONS*, p. 76.

25. Joel N. Shurkin, "Physicists' Secret Organization Seeks Ideas
for Nation's Security," *Chicago Tribune*, February 4, 1973, http://
archives.chicagotribune.com/1973/02/04/page/3/article/physicists
-secret-organization-seeks-ideas-for-nations-security.

26. Zuoyue Wang, *In Sputnik's Shadow: The President's Science Advisory
Committee and Cold War America* (New Brunswick, NJ: Rutgers University
Press, 2008).

27. Finkbeiner, *JASONS*, p. 78.

28. Edward J. Drea, *McNamara, Clifford, and the Burdens of Vietnam, 1965–
1969* (Washington, DC: Office of the Secretary of Defense, 2011).

29. Gell-Mann, interview with Daniel Ford.

30. Bridger, "Best Intentions."

31. Finkbeiner, *JASONS*, p. 82.

32. Ibid., p. 83.

33. Garwin, interview with the author.

34. Finkbeiner, *JASONS*, p. 98.

35. Ibid., p. 85.

36. Ibid., p. 86.

37. Garwin, interview with the author.

38. Finkbeiner, *JASONS*, p. 104.

39. Joel N. Shurkin, "The Secret War Over Bombing," *Philadelphia
Inquirer*, February 4, 1973.

40. Ibid.

41. Finkbeiner, *JASONS*, p. 103.

42. Wang, *Sputnik's Shadow*.

43. Sidney Drell, in an interview with the author, May 26, 2016.

44. Ibid.

45. Garwin, interview with the author.

46. Finkbeiner, *JASONS*.

47. Ibid., p. 112.

CHAPTER TEN: SUPER SONIC TRANSPORT

1. Tom D. Crouch, *Wings: A History of Aviation from Kites to the Space Age* (New York: W. W. Norton, 2003), p. 625.

2. Jason Rabinowitz, "JFK Furious Over Pan Am Concorde Order in Declassified Phone Calls," *Airways News*, February 12, 2014, http://airwaysnews.com/blog/2014/01/09/jfk-furious-pan-concorde-order -declassified-phone-calls/ (accessed July 21, 2016).

3. Ibid.

4. Ibid.

5. Ibid.

6. Richard L. Garwin, in an interview with the author, February 9, 2016.

7. Ibid.

8. Ibid.

9. Ibid.

10. Ibid.

11. Ann Finkbeiner, *The JASONS: The Secret History of Science's Postwar Elite* (New York: Penguin, 2006), p. 173.

12. David Kestenbaum, "An Evening With Richard Garwin," *National Public Radio* (Jan.10, 2016).

13. Peter Rogers, "Concorde Booms and the Mysterious East Coast Noises," *Acoustics Today*, 11 (Spring 2015).

14. T. A. Heppenheimer, *The Space Shuttle Decision: NASA's Search for a Reusable Space Vehicle* (Washington, DC: NASA, 1999), p. 309, https://ntrs .nasa.gov/archive/nasa/casi.ntrs.nasa.gov/19990056590.pdf (accessed November 8, 2016).

15. Ibid.

16. Ibid., p. 310.

17. Richard L. Garwin, in an interview with the author, February 2, 2016.

18. Richard L. Garwin, testimony, US Senate Appropriations Committee, August 28, 1976.

19. Kestenbaum, "An Evening With Richard Garwin."

20. Garwin, testimony.

21. Zuoyue Wang, *In Sputnik's Shadow: The President's Science Advisory Committee and Cold War America* (New Brunswick, NJ: Rutgers University Press, 2008).

22. Ibid.

23. Richard L. Garwin, in an interview with the author, March 24, 2016.

24. Ibid.

25. Ibid.

26. Ibid.

27. Ibid.

CHAPTER ELEVEN: OFFENSE

1. Richard L. Garwin, "A Cruise Missile System for Europe," *Sack Memorial Lecture* (lecture, Cornell University, Ithaca, NY, September 26, 1978).

2. George C. Wilson, "Cruise Missile Test Failures Point Up Sub Plan's Problems," *Washington Post*, July 27, 1978, https://www.washingtonpost.com/archive/politics/1978/07/27/cruise-missile-test-failures-point-up-sub-plans-problems/09c883e0-a26e-451d-b3d8-7758f4d8fabe/.

3. Garwin, "Cruise Missile System."

4. Ibid.

5. Richard L. Garwin, testimony, US House Armed Services Committee, April 17, 1975.

6. Richard L. Garwin, in an interview with the author, January 29, 2016.

7. Ibid.

8. Kris Osborn, "Air Force Begins Massive B-1B Overhaul," *Defensetech*,

February 21, 2014, http://www.defensetech.org/2014/02/21/air-force -begins-massive-b-1b-overhaul/ (accessed November 13, 2016).

9. Richard L. Garwin, in an interview with the author, May 13, 2016.

10. Ibid.

11. Tom Garwin, in an interview with the author, June 8, 2016.

12. Richard L. Garwin, in an interview with the author, September 9, 2015.

13. Ibid.

14. Ibid.

15. Dwight Eisenhower, "Farewell Radio and Television Address to the American People," January 17, 1961, transcript, President's Office, White House, https://www.eisenhower.archives.gov/all_about_ike/speeches/farewell _address.pdf (accessed November 1, 2016).

16. Garwin, interview with the author, January 29, 2016; Richard L. Garwin, in an interview with the author, February 9, 2016.

17. Garwin, interview with the author, February 9, 2016.

18. Ann Finkbeiner, *The JASONS: The Secret History of Science's Postwar Elite* (New York, NY: Penguin, 2006).

19. Garwin, interview with the author, February 9, 2016.

20. Sidney Drell, in an interview with the author, May 26, 2016.

21. Ibid.

22. Richard L. Garwin, in a personal document, "Sid Drell and National Security," undated.

23. Drell, interview with the author.

24. Ibid.

25. Richard L. Garwin, in an interview with the author, January 25, 2016.

26. William Perry, in an interview with the author, June 13, 2016.

27. Garwin, interview with the author, February 9, 2016.

28. Tom Garwin, in an interview with the author.

29. Drell, interview with the author.

30. Richard L. Garwin, in an interview with the author, February 2, 2016.

31. Theodore Postol, in an interview with the author, July 1, 2016.

32. Theodore Postol, in a letter to Richard Garwin, June 24, 2015.

33. Garwin, interview with the author, February 2, 2016.

34. Ibid.

35. Ibid.

36. Sidney D. Drell and Richard L. Garwin, MIT, "Basing the MX Missile: A Better Idea." *Technology Review* (May/June 1981).

37. Garwin, interview with the author, January 29, 2016.

38. Richard L. Garwin, in a letter to Thomas T. Schelling, "Comments on MX," June 17, 1976.

39. Office of Technology Assessment, "MX Missile Basing," (September 1981), http://ota.fas.org/reports/8116.pdf.

40. Heritage Foundation, "One Cheer for the Scowcroft Commission," April 20, 1983.

41. Richard L. Garwin, "The Proper Strategic Mix in Lieu of Dense-Packed Mx's," *New York Times*, January 24, 1983, http://www.nytimes.com/1983/01/24/opinion/l-the-proper-strategic-mix-in-lieu-of-dense-packed-mx-s-180431.html.

CHAPTER TWELVE: THE GREAT GAP

1. Richard L. Garwin, in an e-mail to the author, November 2016.

2. Greg Canavan, in an interview with the author, May 25, 2016.

3. Ibid.

4. Arms Control Association, "The Anti-Ballistic Missile (ABM) Treaty at a Glance," (January 1, 2003), https://www.armscontrol.org/factsheets/abmtreaty (accessed July 25, 2016).

5. Richard L. Garwin, in an interview with the author, June 16, 2016.

6. Missy Ryan, "Pentagon to Deploy Anti-Missile System in South Korea," *Washington Post*, July 7, 2016, https://www.washingtonpost.com/news/checkpoint/wp/2016/07/07/pentagon-to-deploy-anti-missile-system-in-south-korea/.

7. Richard L. Garwin and Hans A. Bethe, "Anti-Ballistic-Missile Systems," *Scientific American*, 218 no. 3 (March 1, 1968): 21–31.

8. Richard L. Garwin, in an e-mail to the author, June 14, 2016.

9. Garwin and Bethe, "Anti-Ballistic-Missile Systems."

10. Garwin, interview with the author June 16, 2016.

11. Richard L. Garwin, "Missile Defense: NMD, TMD, EMD . . .," *Aspen Institute*, Berlin, Germany, March 1, 2001, http://fas.org/rlg/010301-aspen.htm (accessed July 25, 2016).

12. Canavan, interview with the author.

13. Garwin, interview with the author June 16, 2016.

14. Ibid.

15. Ibid.

16. Ibid.

17. Garwin and Bethe, "Anti-Ballistic-Missile Systems."

18. Ibid.

19. Ibid.

20. Garwin, interview with the author June 16, 2016.

21. Richard L. Garwin, "Scientist, Citizen, and Government: Ethics in Action or Ethics Inaction," *Illinois Mathematics and Science Academy*, Aurora, IL, May 4, 1993, http://fas.org/rlg/930504-imsa.htm (accessed July 21, 2016).

22. Richard L. Garwin, "Living with Nuclear Weapons: Sixty Years and Counting," American Philosophical Society (lecture, Philadelphia, PA, April 2005).

23. Ibid.

24. Ibid.

25. Ibid.

26. Garwin, interview with the author, June 16, 2016.

27. Ibid.

28. "CIA Evidence of an Israeli Nuclear Test," *Guardian*, August 12, 2010, https://www.theguardian.com/world/2010/aug/13/evidence-of -israeli-nuclear-test.

29. Leonard Weiss, "Flash from the Past: Why an Apparent Israeli Nuclear Test in 1979 Matters Today," *Bulletin of the Atomic Scientists*, September 8, 2015, http://thebulletin.org/flash-past-why-apparent-israeli -nuclear-test-1979-matters-today8734 (accessed July 25, 2016).

30. Ibid.

31. Ibid.

32. Ibid.

33. Ibid.

34. Ibid.

35. Luis Alvarez summary, quoted in Carey Sublette, "Report on the 1979 Vela Incident," *Nuclear Weapon Archive*, September 1, 2001, http:// nuclearweaponarchive.org/Safrica/Vela.html (accessed November 15, 2016).

36. Jeffrey T. Richelson, *Spying on the Bomb: American Nuclear Intelligence from Nazi Germany to Iran and North Korea* (New York: W. W. Norton, 2007), p. 296.

37. Richard L. Garwin, in an interview with the author, June 2, 2016.

38. Ibid.

CHAPTER THIRTEEN: TREATY

1. "The Treaty on the Non-Proliferation of Nuclear Weapons," *UN.org*, http://www.un.org/en/conf/npt/2005/npttreaty.html (accessed November 13, 2016).

2. Richard L. Garwin, in an interview with the author, June 24, 2016.

3. Ibid.

4. Ibid.

5. Ibid.

6. Richard L. Garwin, "*60 Minutes* on Particle Beam Weapons," *Bulletin of the Atomic Scientists* (February 1979).

7. Garwin, interview with the author, June 24, 2016.

8. Ibid.

9. Richard L. Garwin, unpublished manuscript.

10. Garwin, interview with the author, June 24, 2016.

11. Ibid.

12. Ibid.

13. Ibid.

14. Richard L. Garwin, in an interview with the author, June 28, 2016.

15. Garwin, "*60 Minutes* on Particle Beam Weapons."

CHAPTER FOURTEEN: STAR WARS

1. Francis Fitzgerald, *Way Out There in the Blue: Star Wars and the End of the Cold War* (New York: Simon & Schuster, 2000), pp. 22–23.

2. This and the next several paragraphs on Postol from Theodore Postol, in an interview with the author, July 1, 2016.

3. Ibid.

4. Fitzgerald, *Way Out There in the Blue*, p. 163.

5. Richard L. Garwin, in an interview with the author, May 3, 2016.

6. Fitzgerald, *Way Out There in the Blue*, p. 38.

7. Atomic Archive, "Cold War: A Brief History, Reagan's Star Wars," http://www.atomicarchive.com/History/coldwar/page20.shtml (accessed July 25, 2016).

8. Science x Network, "30 Years and Counting, the X-Ray Laser Lives On," *Phys.org*, April 15, 2015, http://phys.org/news/2015-04-years-x-ray-laser.html (accessed July 25, 2016).

9. Garwin, interview with the author, May 3, 2016.

10. Ibid.

11. Edward A. Frieman, John Cornwall, and Mal Ruderman, in an interview with Daniel Ford, June 2004.

12. Ibid.

13. Ibid.

14. Greg Canavan, in an interview with the author, May 25, 2016.

15. Richard L. Garwin, in an interview with the author, June 27, 2016.

16. Canavan, interview with the author, May 25, 2016.

17. Frieman, Cornwall, and Ruderman, interview with Daniel Ford.

18. Ibid.

19. Garwin, interview with the author, May 3, 2016.

20. John Broder, "'Star Wars' First Phase Cost Put at $170 Billion: System Would Intercept Only 16% of Soviet Missiles, Report of 3 Senate Democrats Says," *Los Angeles Times*, June 12, 1988, http://articles.latimes.com/1988-06-12/news/mn-7383_1_star-wars.

21. Garwin, interview with the author, May 3, 2016.

22. "Is the Strategic Defense Initiative in the National Interest," C-SPAN, November 17, 1987, https://www.c-span.org/video/?532-1/sdi-debate (accessed November 13, 2016).

23. Ibid.

24. Ibid.

25. Richard L. Garwin, "*60 Minutes* on Particle Beam Weapons," *Bulletin of the Atomic Scientists* (February 1979).

26. Richard L. Garwin, in an interview with the author, May 13, 2016.

27. Frieman, Cornwall, and Ruderman, interview with Daniel Ford.

CHAPTER FIFTEEN: GRAVITY

1. Richard L. Garwin, in an interview with the author, March 8, 2016.
2. Ibid.
3. Walter Munk and Ed Frieman, in an interview with Daniel Ford, July 2, 2004.
4. Richard L. Garwin, in an interview with the author, March 8, 2016.
5. Garwin, interview with the author, March 8, 2016.
6. Richard L. Garwin and James L. Levine, "Single Gravity-Wave Detector Results Contrasted with Previous Coincidence Detections," *Physical Review Letters*, 31 (July 16, 1973).
7. Richard L. Garwin, in an interview with the author, May 6, 2016.
8. Ibid.
9. Freeman Dyson, in an interview with the author, February 20, 2014.
10. Garwin, interview with the author, May 6, 2016.
11. Jennifer Chu, "A 40-Year Quest to Prove Einstein Right," *MIT Technology Review*, April 26, 2016, https://www.technologyreview.com/s/601147/a-40-year-quest-to-prove-einstein-right/ (accessed July 25, 2016).
12. Ibid.
13. "Gravitational Waves Spotted Again," *CNRS Press Releases*, June 15, 2016, http://www2.cnrs.fr/en/2769.htm (accessed October 27, 2016).
14. Richard L. Garwin, in an interview with the author, June 2, 2016.
15. Dino A. Brugioni, *Eyes in the Sky: Eisenhower, the CIA, and Cold War Aerial Espionage* (Annapolis, MD: Naval Institute Press, 2010).
16. Garwin, interview with the author, June 2, 2016.
17. Ibid.
18. Kevin C. Ruffner, ed., *Corona: America's First Satellite Program* (Washington, DC: CIA, 1995).
19. Ibid.
20. Garwin, interview with the author, June 2, 2016.

CHAPTER SIXTEEN: HEALTH, PANDEMICS

1. Richard L. Garwin, "Lessons from the 2009 Swine Flu Pandemic, Avian Flu, and their Contributions to the Conquest of Induced and Natural

Pandemics," *Erice International Seminars*, August 22, 2010, http://fas.org/rlg/2010%20Erice%20Learning%20from%20Pandemics_1.pdf (accessed July 25, 2016).

2. Ibid.

3. Stephen S. Morse, Richard L. Garwin, and Paula J. Olsiewski, "Next Flu Pandemic: Until the Vaccine Arrives," *Science*, 314, no. 5801 (November 10, 2006): 929.

4. Richard L. Garwin, in an interview with the author, March 14, 2016.

5. Morse, Garwin, Olsiewski, "Next Flu Pandemic."

6. Garwin, interview with the author, March 14, 2016.

7. Richard L. Garwin, in an interview with the author, April 14, 2016.

8. For this and remainder of quotes in chapter: Richard L. Garwin, "Impact of Information-Handling Systems on Quality and Access to Health Care," *Public Health Reports*, 83, no. 5 (May 1968): 346–51.

CHAPTER SEVENTEEN: FAR OUT

1. "HSCA Final Assassinations Report," *History Matters Archive*, http://www.history-matters.com/archive/jfk/hsca/report/html/HSCA_Report_0005a.htm (accessed November 14, 2016).

2. This quote and next several quotes from Richard L. Garwin, in an interview with the author, July 1, 2016.

3. National Academy of Sciences, *Report of the Committee on Ballistic Acoustics* (Washington, DC: National Academies Press, 1982).

4. Richard L. Garwin, "Proposal for an International Air, Sea, and Space Traffic Control Using Radio Beacons on the Vehicles," *IBM Watson Laboratory*, July 20, 1958, http://fas.org/rlg/072058pias.pdf (accessed July 25, 2016).

5. Ibid.

6. Ibid.

7. Ibid.

8. Richard L. Garwin, "Fun with Muons, GPS, Radar, etc.," *Lee Historical Lecture* (lecture, Harvard University, Cambridge, MA, March 18, 2003), http://fas.org/rlg/FunWithMuons051607b.pdf (July 22, 2016).

9. Richard L. Garwin, in an interview with the author, undated.

10. Ibid.

11. Ibid.

12. Ibid.

13. Paul Gilster, "A Solar Sail Manifesto," *Centauri Dreams*, April 30, 2009, http://www.centauri-dreams.org/?p=7545 (accessed November 15, 2016).

14. Paul Gilster, "SF Influences: A Solar Sail Theory," *Centauri Dreams*, March 18, 2014, http://www.centauri-dreams.org/?p=30223 (accessed July 25, 2016); Colin R. McInnes, *Solar Sailing: Technology, Dynamics and Mission Applications* (New York: Springer Praxis, 1999), p. 2; Richard L. Garwin, interview with the author, January 11, 2015.

15. Richard L. Garwin, "Solar Sailing: A Practical Method of Propulsion within the Solar System," *Jet Propulsion* (March 1958).

16. Ibid.

17. Dennis Overbye, "Setting Sail Not Space, Propelled by Sunshine," *New York Times*, November 9, 2009, http://www.nytimes.com/2009/11/10/science/space/10solar.html.

18. Planetary Society, "The Mission: Sailing in Space," *Planetary Society*, http://sail.planetary.org (accessed December 30, 2015).

19. Richard L. Garwin, in an interview with the author, September 9, 2015.

20. Richard L. Garwin, in an interview with the author, April 14, 2016.

21. Paul Michael Grant, "Superconducting Lines for the Transmission of Electrical Power over Great Distances: Garwin-Matisoo Revisited 40 Years Later," *IEEE Transactions*, 17, no. 2 (June 2007): 1641–1647.

22. Richard L. Garwin and J. Matisoo, "Superconducting Lines for the Transmission of Large Amounts of Electrical Power over Great Distances," *Proceedings of the IEEE*, 55 (April 1967): 538.

23. Richard L. Garwin, in an interview with the author, May 3, 2016.

24. Richard L. Garwin, in an interview with the author, April 22, 2016.

25. Ibid.

26. Ibid.

27. Richard L. Garwin, in a letter to John J. Martin, October 30, 1978.

28. Ibid.

29. Richard L. Garwin, in a letter to Harold Brown, February 28, 1980.

30. Ibid.

31. Richard L. Garwin, in a letter to Zbigniew Brzezinski, January 28, 1980.

32. Joshua Lederberg, in a letter to Richard Garwin, April 14, 1993.

CHAPTER EIGHTEEN: RUMPLED

1. Jeffrey Garwin, in an interview with Daniel Ford, January 16, 2011.

2. Lois E. Beckett, "A Brassy Move," *Harvard Crimson*, December 11, 2006, http://www.thecrimson.com/article/2006/12/11/a-brassy-move-the-sound-of/ (accessed November 15, 2016).

3. Jeffrey Garwin, interview with Daniel Ford.

4. Lois Garwin, in an interview with Daniel Ford, July 1, 2004.

5. Jeffrey Garwin, interview with Daniel Ford.

6. Lois Garwin, interview with Daniel Ford.

7. William Perry, in an interview with the author, June 13, 2016.

8. Jonathan Katz, in an interview with Daniel Ford, September 3, 2007.

9. Marvin "Murph" Goldberger, in an interview with Daniel Ford, December 20, 2012.

10. Katz, interview with Daniel Ford.

11. Edward A. Frieman, John Cornwall, and Mal Ruderman, in an interview with Daniel Ford, June 2004.

12. Ibid.

13. Walter Munk and Ed Frieman, in an interview with Daniel Ford, July 2, 2004.

14. Frieman, Cornwall, and Ruderman, interview with Daniel Ford.

15. Ibid.

16. Jack Ruina, in an interview with Daniel Ford, December 2004.

17. Ann Finkbeiner, *The JASONS: The Secret History of Science's Postwar Elite* (New York, NY: Penguin, 2006).

18. Ruina, interview with Daniel Ford.

19. Freeman Dyson, in an interview with the author, February 20, 2014.

20. Wolfgang "Pief" Panofsky, in an interview with Daniel Ford, December 7, 2004.

21. Finkbeiner, *JASONS*, p. 176.

22. Theodore Postol, in an interview with the author, July 1, 2016.

23. Munk and Frieman, interview with Daniel Ford.

24. Finkbeiner, *JASONS*, p. 177.

25. Richard L. Garwin, interview by Finn Aaserud, *Niels Bohr Library & Archives*, AIP, June 24, 1991, https://www.aip.org/history-programs/niels -bohr-library/oral-histories/5075 (accessed December 5, 2016).

26. Richard L. Garwin, in an interview with Daniel Ford, May 17, 2006.

27. Richard L. Garwin, in personal document, "Sid Drell and National Security," undated.

28. John Parmentola, in an interview with the author, undated.

29. Joel N. Shurkin, Broken Genius: The Rise and Fall of William Shockley, Creator of the Electronic Age (New York: Macmillan, 2006).

30. William J. Broad, "29 US Scientists Praise Iran Nuclear Deal in Letter to Obama," *New York Times*, August 8, 2015, http://www.nytimes.com/2015/08/09/world/29-us-scientists-praise-iran-nuclear-deal-in-letter-to-obama.html.

31. Richard L. Garwin, in an interview with the author, June 19, 2016.

32. Laura Garwin, in an interview with the author, April 8, 2016.

33. Tom Garwin, in an interview with the author, June 8, 2016.

34. Laura Garwin, interview with the author.

35. Richard L. Garwin, "*60 Minutes* on Particle Beam Weapons," *Bulletin of the Atomic Scientists* (February 1979).

36. Ibid.

37. Ibid.

38. Richard L. Garwin, in an interview with the author, September 20, 2015.

39. Ibid.

CHAPTER NINETEEN: DECLINE OF INFLUENCE

1. William Perry, in an interview with the author, June 13, 2016.

2. Theodore Postol, in an interview with the author, July 1, 2016.

3. Dwight Eisenhower, "Farewell Radio and Television Address to the American People," January 17, 1961, transcript, President's Office, White House, https://www.eisenhower.archives.gov/all_about_ike/speeches/farewell _address.pdf (accessed November 1, 2016).

4. Perry, interview with the author.

5. Frank von Hippel, in an interview with the author, July 1, 2016.

6. Richard L. Garwin, "Scientist, Citizen, and Government: Ethics in Action (or Ethics Inaction)," *Illinois Mathematics and Science Academy*, Aurora, Illinois, May 4, 1993, http://fas.org/rlg/930504-imsa.htm (accessed July 21, 2016).

7. Von Hippel, interview with the author.

8. Ibid.

9. Postol, interview with the author.

10. James C. Kaufman, "The American Idol Effect: Why We're Not Too Good at Judging Our Own Creativity," *Psychology Today*, February 26, 2010, https://www.psychologytoday.com/blog/and-all-jazz/201002/the-american-idol-effect-why-were-not-too-good-judging-our-own-creativity.

11. Postol, interview with the author.

12. Ibid.

13. Richard L. Garwin, in an interview with the author, May 15, 2016.

14. Ibid.

15. Ibid.

16. Richard L. Garwin, "Personal Experience in Advising the US Government: 1956–2007," *Cornell-PRIF Conference on Science Advising and International Security* (lecture, Cornell University, Ithaca, NY, February 23–24, 2007), http://fas.org/rlg/022407_Personal_Experience_in_Advising_p2.pdf (accessed July 22, 2016).

17. Ibid.

18. Ibid.

AFTERWORD

1. Richard L. Garwin, in an interview with the author, May 15, 2016.

2. "President Obama Names Recipients of the Presidential Medal of Freedom," Office of the Press Secretary, November 16, 2016, https://www.whitehouse.gov/the-press-office/2016/11/16/president-obama-names-recipients-presidential-medal-freedom (accessed November 17, 2016).

INDEX